THE FRANZIS ARDUINO™ TUTORIAL KIT

ORIGINAL ARDUINO UNO AND 20 OTHER COMPONENTS FOR 65 PROJECTS

© 2016 Franzis Verlag GmbH, Richard-Reitzner-Allee 2, 85540 Haar
ISBN: 978-3-645-65279-7

Translation and DTP: G&U Language & Publishing Services GmbH
Layout: bora-dtp

All circuits and programs depicted in this book are developed and tested with utmost care. Nonetheless, it is not possible to rule out all errors in the book or in the software. Publisher and author are only liable in case of intent or gross negligence according to legal regulation. Beyond that, publisher and author are only liable according to the law on product liability concerning hazards to life, body, and health and the culpable violation of essential contractual obligations. The damage claim for the violation of essential contractual obligations is limited to the contract-specific, predictable damage, unless in cases of mandatory liability according to the law on product liability.

Dear customers!
This product was developed in compliance with the applicable European directives and therefore carries the CE mark. Its authorized use is described in the instructions enclosed with it. In the event of non-conforming use or modification of the product, you will be solely responsible for complying with the applicable regulations. You should therefore take care to assemble the circuits as described in the instructions. The product may only be passed on along with the instruction and this note.

Waste electrical products should not be disposed of with household waste. Please recycle where facilities exist. Check with your local authority or retailer for recycling advice.

All rights reserved, including those of reprinting, reproduction and storage in electronical media. No part may be reproduced and distributed on paper, on storage media, or in the Internet, especially as PDF, without the publisher's prior written permission. Any attempt may be prosecuted. Hardware and software product names, company names, and company logos mentioned in this book are generally registered trademarks and have to be considered as such. For product names, the publisher uses mainly the spelling of the manufacturer.

Arduino™ is a registered trademark of Arduino LLC and the associated companies.

Table of Contents

Preface — 7

1 Microcontroller Basics — 10

- 1.1 | Measuring — 12
- 1.2 | Controlling — 12
- 1.3 | Controlling with continuous adjustment — 13
- 1.4 | Design and mode of operation — 14
- 1.5 | Programming a Microcontroller — 16

2 A Survey of Available Arduino Boards — 20

- 2.1 | Arduino Mega — 22
- 2.2 | Arduino Uno — 23
- 2.3 | Arduino Leonardo — 24
- 2.4 | Arduino Ethernet — 26
- 2.5 | ArduPilot — 27
- 2.6 | LilyPad — 28
- 2.7 | USB adapter — 29

3 Arduino Shields — 30

- 3.1 | Arduino ProtoShield — 31
- 3.2 | Ardumoto — 32
- 3.3 | TellyMate — 33
- 3.4 | XBee radio frequency modules — 35
- 3.5 | Ethernet shield — 37

4   Components in the Tutorial Kit   38

     4.1 | A survey of the components   39
     4.2 | Arduino Uno   40
     4.3 | Ports and LEDs of the Arduino Uno   41
     4.4 | Power supply   44
     4.5 | Reset button   44
     4.6 | ISP port   44
     4.7 | Safety notes   45

5   Use of the Components   46

     5.1 | Jump wire   47
     5.2 | Breadboard   48
     5.3 | Push-buttons   49
     5.4 | Resistors   49
     5.5 | Capacitors   54
     5.6 | Piezo buzzer   56
     5.7 | LEDs   56
     5.8 | Diode   58
     5.9 | Transistors   59

6   Installation of the Programming Environment   62

     6.1 | Installation on Windows   63
     6.2 | Installation on Mac OS X   71
     6.3 | Installation on Linux   72

7   Arduino Programming Environment   74

Table of Contents

8 Your First Arduino Program — 78
8.1 | What did we do? — 82

9 Arduino Programming Basics — 86
9.1 | Bits and Bytes — 87
9.2 | Structure of a Program — 88
9.3 | Our second Arduino Program — 92
9.4 | Getting Started with Arduino Programming — 95

10 More Experiments with the Arduino — 178
10.1 | LED dimmer — 180
10.2 | Soft flasher — 184
10.3 | Debouncing buttons — 189
10.4 | A simple switch-on delay — 195
10.5 | A simple switch-off delay — 197
10.6 | LEDs — 199
10.7 | Switching large consumers — 202
10.8 | Using the PWM Pins as DAC — 206
10.9 | Music's in the air — 212
10.10 | Romantic Candlelight, Courtesy of the Microcontroller — 217
10.11 | Surveillance at the Exit for Staff Members — 220
10.12 | An Arduino Clock — 223
10.13 | School Bell Program — 225
10.14 | Keypad Lock — 230
10.15 | Capacitance meter with auto-range function — 235

10.16 | Reading potentiometers and trimmers the professional way	239
10.17 | State Machines	242
10.18 | 6-channel voltmeter	247
10.19 | Programming Your Own Voltage Plotter	250
10.20 | Arduino Storage Oscilloscope	253
10.21 | StampPlot: a professional data logger – free of charge!	255
10.22 | Controlling the Arduino Pins via the Arduino Ports Program	261
10.23 | Temperature Switch	264

11 The Fritzing Program — 268

12 The Processing Program — 270

13 Appendix — 274

13.1 | Electrical quantities	275
13.2 | ASCII Table	277

Preface

With many microcontroller systems, you have to work through countless data sheets that are incomprehensible for beginners. The programming interfaces are very complex and devised for professional developers with years of experience in programming microcontrollers. Thus, the access to the world of microcontrollers is unnecessarily made complicated.

The Arduino system, however, is an easily comprehensible open-source platform that is easy to understand. It is based on a microcontroller board with an Atmel AVR controller and a simple programming environment. For the human-machine interaction, you can attach a variety of analog and digital sensors that capture ambient quantities and pass the data to the microcontroller where they are processed. The program causes the creation of new analog or digital output data. There is no limit to the creativity of the developer. Whether you want to build a control system for your home or a beautiful LED lamp with changing colours: The Arduino allows even beginners from another background to write functional programs and to put their own ideas into practise.

The smooth cooperation of hardware and software is the basis for »physical computing« – the linking of the real world to the bits-and-bytes world of the microcontroller.

This tutorial kit conveys the basics of electronics and Arduino programming and shows in a plain way how to implement your own ideas.

Ulli Sommer

The CD in the Tutorial Kit

This tutorial kit contains a CD with several programs, tools, data sheets, and examples. The CD is intended to help you in working with this book. All examples in this book are contained on the CD as well.

The contents of the CD

The contents of the CD

- Arduino IDE (Integrated Development Environment)
- Sample program code
- Several tools
- Data sheets
- Circuit diagrams

GPL (General Public Licence)

You can share your own programs on the internet with other users. The sample programs are provided under the open-source GPL licence (General Public Licence). This means that you have the right to modify, publish, and share the programs according to the conditions of the GPL, provided that you make them available under the same licence terms.

System Requirements

- Windows XP (32- or 64-bit) or newer; or:
- Linux (32- or 64-bit); or:
- Mac OS X.
- CD drive
- Java

More information can be found on the following websites:

Further reading

- www.arduino.cc
- www.fritzing.org
- www.processing.org

Updates and Support

The Arduino IDE is continually developed further. You can download any updates free of charges at the following website:

hptt://arduino.cc

Warning! Eye protection in handling LEDs

Never look directly to an LED at a short distance! This could damage your retina! This is especially true for bright LEDs in a clear housing und even more for Power LEDs. The perceived brightness of white, blue, purple, and ultraviolet LEDs gives a false impression of the real danger for your eyes. Always exercise extreme caution when using convergent lenses. Operate any LEDs according to the instructions, and never use higher currents.

TURN ON YOUR CREATIVITY
FRANZIS
ARDUINO

MICRO-
CONTROLLER
BASICS

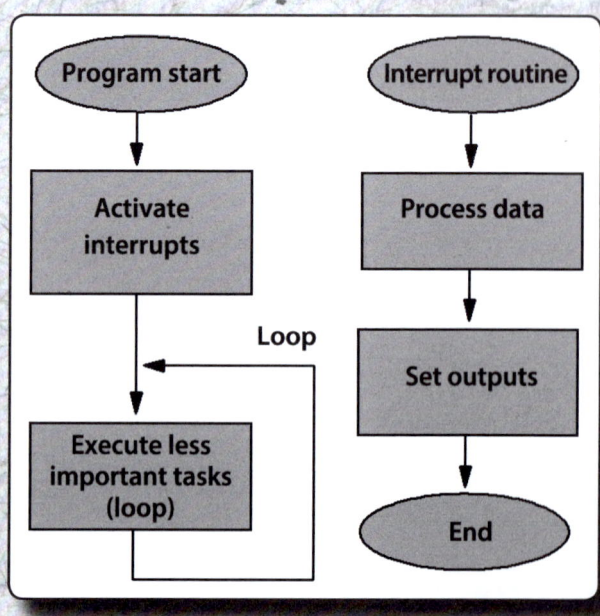

1

Before we go into the details of the Arduino, we want to get a general idea about microcontrollers. Microcontrollers are mainly used in the area of measurement and control. A microcontroller system provides the advantage of measuring and controlling physical quantities in an energy- and cost-efficient way and with a very small footprint. Based on these measurements, it executes its actions. Basically, any task you want to undertake with your Arduino is a measurement and control task.

Microcontroller systems work in the smallest space

... THE ARDUINO'S MAIN TASKS: MEASUREMENT AND CONTROL!

1.1 | Measuring

The Arduino's main tasks: measurement and control!

»Measuring« generally means to capture physical input values of push-buttons, switches, light sensors, pressure sensors, motion detectors, photo-electric guards, potentiometers, etc. These values are fed into the controller via the digital or analog inputs. The Arduino can read signals with a voltage level between 0 and 5 V on the digital inputs. The analog inputs can evaluate 0 to 5 V with a resolution of 10 bit, where 0 V equals 0 and 5 V equals 1023. This value can subsequently be interpreted by our software, e.g. to test a battery's (AA) voltage. For higher voltages on the inputs (no matter whether digital or analog), we need a potentiometer that reduces the input voltage to the maximum input voltage for Arduino pins. This will be discussed in more detail later.

1.2 | Controlling

Controlling means to respond to an input value

»Control« means to respond to an input value. Let's take an electric iron as a simple example: The temperature is controlled by means of a bimetallic switch. When the surface is colder than the preset temperature, it will be heated until the desired temperature is reached. When the temperature drops under a specified value, heating will recommence until the temperature is right again. There is a little »window« between »heating on« and »heating off«, the so-called hysteresis. This is a safeguard against switching the heating on and off too rapidly and thereby wearing out the mechanical contacts or causing a rather »jumpy« control. The bimetallic switch of the microcontroller could be replaced with a temperature sensor. The microcontroller would then capture the value digitally or analogously. The program would then compare the value with the thresholds for »on« and »off« und switch a digital output. This output would control a relay or a transistor for turning the heating on or off as required.

Hysteresis

1.3 | Controlling with continuous adjustment

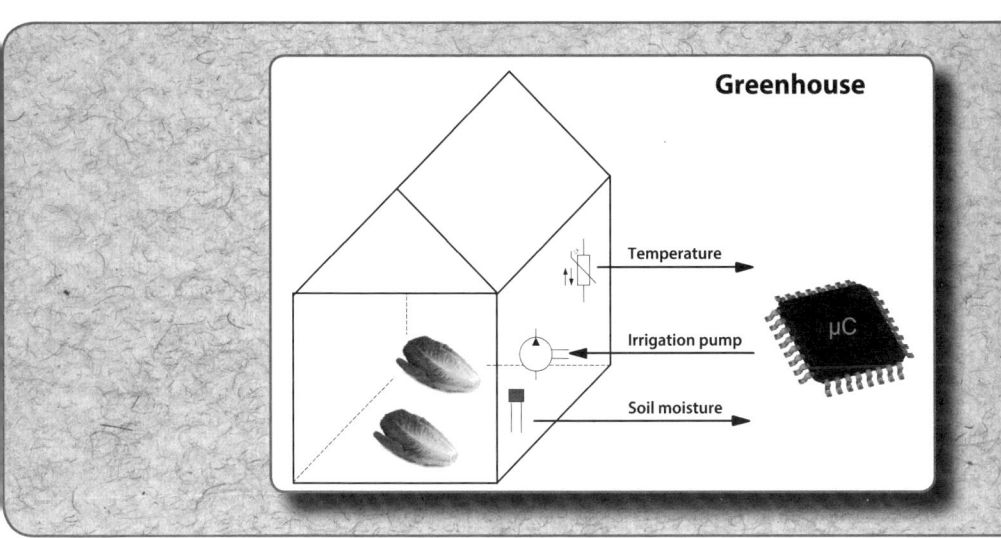

Figure 1.1: Input and output processing in a greenhouse

So far we have only considered controlling with discrete on/off states, but of course, we can also control something by means of continuous adjustment. A good example is the speed control of a car that tries to maintain the preset speed. If one had only the discrete responses »accelerate«, »do nothing«, and »brake«, driving with the speed control would be extremely uncomfortable!

Possible uses for the Arduino – at home and professionally

The possible applications for microcontrollers range from private use (e.g. controlling greenhouses, aquariums, or lights) to the industrial area, where complete production lines can be controlled, maintained, and operated via microcontroller systems. Figure 1.1 shows the typical data processing for controlling the irrigation system of a greenhouse. The microcontroller uses sensors to measure temperature and soil moisture. These measured values are then interpreted by the digital logic of a software in the microcontroller (MC for short) that determines whether the pump is turned on or off.

1.4 | Design and mode of operation

Basic elements of microcontrollers

Any microcontroller is a full-fledged computer despite its tiny size. Like a PC, it consists of some basic electronic components. These basic elements are the CPU, the (main) memory (RAM), the program memory (flash), and the peripherals (see Fig. 1.2).

1.4.1 | The CPU

The CPU is the brain of the microcontroller

The most important component is the central processing unit (CPU), the »brain« of the microcontroller. It processes the instructions and arithmetical operations.

1.4.2 | Main memory and program memory

The main memory is used to store data temporarily

Main memory and program memory are separated logically. Typically, user programs are stored in non-volatile flash memory, the so-called program memory. Depend-

1 Microcontroller Basics

ing on the model of the microcontroller, the size of the program memory can encompass several kilobyte or megabyte. With some systems, it is even possible to extend the program memory by attaching external flash components. The main memory (random-access memory, RAM) is used to store calculation data, measurements, and control factors, in order to ensure quick access to a finite set of data. In general, the RAM is much smaller than the flash memory, but considerably faster. Values in the RAM are created at run-time. In contrast to the flash memory, the storage is volatile: After a restart of the controller, the stored values in the RAM are gone.

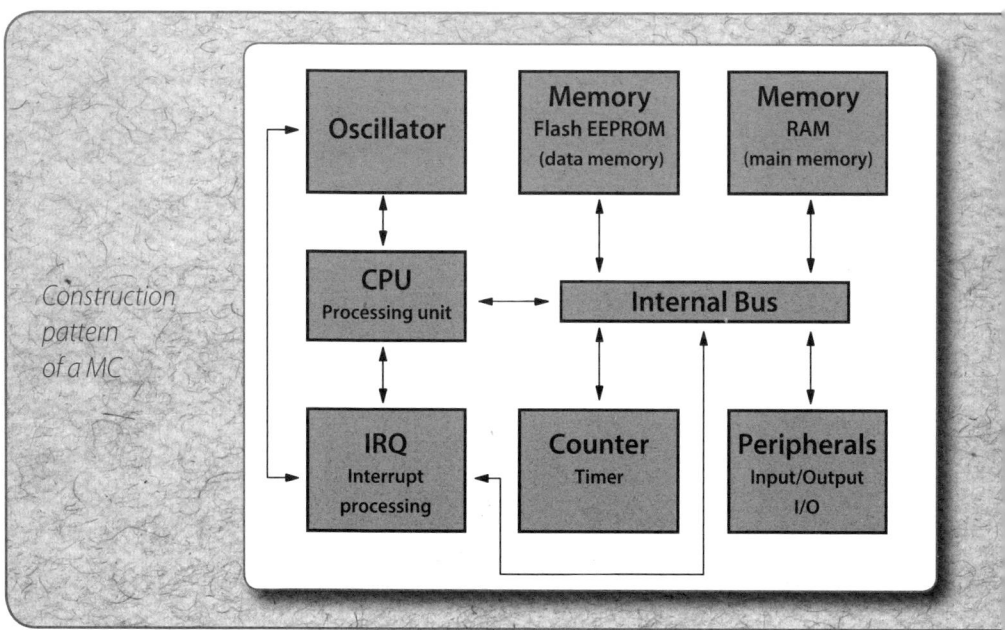

Figure 1.2: General layout of a microcontroller

1.4.3 | Peripherals

The digital inputs and outputs belong to the peripherals

»Peripherals« are all components of a microcontroller that are not covered by the CPU and the memory elements, especially those that interface the outside world, like the digital inputs and outputs (I/O for short). Most microcontroller boards, including the Arduino, provide a plethora of inputs and outputs with diverse functionality. Among them are digital and analog inputs and outputs (ADC and DAC).

1.5 | Programming a Microcontroller

1.5.1 | What is a program?

Program instructions with a particular purpose are called algorithm

A program is the description of an information processing sequence. In this process, a set of output values is calculated on the basis of a set of variable or constant input values. The output values are either the objective of the information gathering process or are used indirectly to respond to the input values. Apart from proper calculations, a program can include instructions to access the hardware or to control the program flow itself. A program consists of several lines of source code. Every line consists of one or more calculations or control instructions. Not only the instructions themselves, but also their order significantly influences the information processing. The control computer executes the operations corresponding to the instructions in sequential order, one after the other. A series of program instructions with a particular purpose is called algorithm.

1.5.2 | Arduino talks C

C and its variant ANSI-C are programming languages that are not hard to learn. C is an imperative programming language and was developed for the operating system Unix in the early 70s by the computer scientist Dennis Ritchie at Bell Laboratories. Since then, its use is spread all over the world. C is used in many different areas, e.g. for system and application programming. The primary programs of all Unix systems and the kernels of many operating systems are written in C. Many languages like C++, Objective-C, C#, Java, PHP, and Perl are based on the syntax and other traits of C. It is worthwhile to get familiar with C, as this gives you the opportunity to switch to other microcontroller systems later on. For nearly every microcontroller, you can download a free C compiler from the manufacturer. The variant of C used in the Arduino is simpler than the professional C compilers and reduces your work load. Primarily, you are spared the tedious hardware routines because they are integrated as fixed orders within the development environment. Moreover, there is an Arduino library for nearly every piece of hardware. These libraries are easily installed. After that, you can communicate with the new hardware, e.g. a digital pressure sensor.

The programming language C

The Arduino hardware consists solely of easily available standard parts. It is therefore easy to understand their function and to adapt the circuits to your needs or to make extensions. The heart of the Arduino is an ATmega controller of the popular 8-bit AVR family of microcontrollers by Atmel. This is complemented by the power supply and a serial interface. The latter is designed as a

Elements of the Arduino

USB interface in newer Arduino versions. The download of programs and, if needed, the communication between PC and Arduino at run-time (e.g. to issue commands or to read measurements from the Arduino) takes place via this interface.

As Arduino boards are designed for ease of use and all purposes, they are sometimes simply called I/O boards. The Arduino Uno provides 14 digital inputs/outputs, six of which can be used as analog outputs (8-bit PWM). Yet another six inputs can capture analog signals (10-bit ADC). If needed, SPI and I^{2}C are available as further digital communication interfaces. Using these, you can extend the Arduino Uno quickly and simply with other components that utilize the same interface.

Arduino boards with different features

There are several variants of Arduino boards, and in addition to the classic 8-bit Arduino, a 32-bit version with decent calculating power for the most complex tasks has recently been introduced.

The original Arduino boards were manufactured by Smart Projects in Italy. They are low-priced and available via many Internet stores. Meanwhile, you can also find many clones and replicas made by other vendors, as the Arduino is open source hardware. An important supporter of the Arduino project is SparkFun in Boulder, Colorado. The cooperation with a partner in the USA has spawned a series of optimized Arduino boards and a vast number

of sensors and actuators. As an important offshoot, the
LilyPad was developed to tackle the topic of Wearable
Computing.

Most users rely on the Arduino Uno by Smart Projects. *Arduino Uno: the*
This version is included in the tutorial kit at hand. *Arduino in this kit*

TURN ON YOUR CREATIVITY

FRANZIS
ARDUINO

A SURVEY OF AVAILABLE ARDUINO BOARDS

2

Well-known variants of the Arduino are the Arduino Mega boards. They have a more powerful ATmega1280 microcontroller, more memory, more I/O pins, and more functions. The Arduino Mini is substantially smaller than the Mega boards and comes in the DIP24 format. The whole module can be fitted onto a 24-pin DIP socket. The Arduino Pro Mini by SparkFun is nearly identical, but is delivered without the small »legs« (pins at the side). For programming these modules, you need a USB adapter that is attached at the narrow side.

Other Arduino variants

Another variant is the LilyPad board by Leah Buechley (in cooperation with SparkFun). It is Arduino-compatible and purpose-built to be sewn into clothes in order to put a very tight symbiosis of technology and art into practice. The typical round form of a LilyPad board is as striking as the colouring and the circular arrangement of the pins. Many tiny peripheral boards (sensors, LEDs, push-buttons, etc.) extend the LilyPad to a system of »electronics with a sewing machine«.

The LilyPad board and its use in functional wear.

Tip Information about further board versions and accessories can be found on the website of the Arduino project. The product pages of SparkFun Electronics are available at:

http://www.sparkfun.com/categories/103

2.1 | Arduino Mega

Figure 2.1: Arduino Mega

Specifications

Specifications
- ATmega2560 microcontroller
- 16 MHz
- 256 KB flash (8 KB of which are used by the bootloader)
- 8 KB SRAM, 4 KB EEPROM
- 54 digital I/O pins, 15 of which are PWM-capable
- Hardware UARTs
- I^2C interface, SPI
- 16 analog inputs (10-bit)
- USB interface, power supply, bootloader, etc. as with the Arduino Duemilanove
- Size: ca. 101 x 53 x 15 mm

2.2 | Arduino Uno

Figure 2.2: Arduino Uno (SMD edition)

This board is included in our tutorial kit. It is the standard Arduino board, based on the Atmel Atmega328P microcontroller. This cheap und powerful board is arguably the best-selling microcontroller board around the world. It has 14 digital I/Os, six of which can be used as PWM outputs. Furthermore, it comes with six analog inputs, a 16 MHz oscillator for a clock generator, a USB-B socket for programming and data output, a reset button, a power supply socket and an ISP connector for programming by an Atmel programmer. With an external power supply, it can only handle direct current.

What the Arduino in this tutorial kit can do

This version is the immediate successor of the first Arduino boards. It retains the original Arduino standards set up in the beginning.

The main differences are: board with SMD mounting, no FTDI-USB-to-USB adapter chip, but instead an ATmega16U2, and a more powerful microcontroller.

Specifications

Specifications
- ATmega328P
- 16 MHz
- 32 KB flash (0.5 KB of which are used by the bootloader)
- 2 KB SRAM, 1 KB EEPROM
- 14 digital I/O pins, 6 of which are PWM-capable
- 6 analog inputs (10-bit)
- On-board USB interface with Atmel ATmega16U2
- 5 V operating voltage via USB or a potentiometer (7 V to 12 V input voltage)
- Size: ca. 69 x 53 x 15 mm

2.3 | Arduino Leonardo

The Leonardo board and its specifications

The Leonardo board is based on the ATmega32U4 microcontroller. It has 20 digital I/Os, seven of which can be used as PWM outputs. You program the board as usual via the USB connection and the bootloader on the ATmega. The difference to other boards is that it has a micro USB socket instead of the standard USB-B connection. Besides, there is no additional USB chip on this board for communication between PC and microcontroller. This chip is already integrated in the microcontroller.

2 A Survey of Available Arduino Boards

Figure 2.3: Arduino Leonardo

This again allows communication with the PC by means of a human interface device (HDI), i.e. a mouse or a keyboard.

Specifications
- ATmega32U4
- 16 MHz
- Programming via USB
- 5 V technology
- 20 digital I/O pins, 7 of which are PWM-capable
- 12 analog inputs (10-bit)
- 32 KB flash (4 KB of which are used by the bootloader)
- 2 KB SRAM
- Output current per I/O pin max. 40 mA
- Supply voltage 7 V to 12 V
- Size: ca. 69 x 53 x 15 mm

2.4 | Arduino Ethernet

Figure 2.4: Arduino Ethernet

Arduino Ethernet and the differences to other Arduino boards

The Arduino Ethernet board is based on an Atmel ATmega328 microcontroller. It has 14 I/Os and six analog inputs. In addition, it comes with a 16 MHz oscillator, an RJ45 connector, a power supply socket, an ICSP (ISP) connector and a reset button. Beyond that, you can solder in a power-over-Ethernet board.

The big difference between this and all the other Arduino boards is that here, an Ethernet connection instead of the usual USB socket is provided for programming. The Ethernet communication is carried out by the on-board Wiznet Ethernet Controller that is attached to the ATmega328. The microSD card reader can be used to store and read data and web pages.

Specifications
- ATmega328
- 16 MHz
- Programming via Ethernet RJ45
- 5 V technology
- 14 digital I/O pins, 4 of which are PWM-capable
- 6 analog inputs (10-bit)
- 32 KB Flash (0.5 KB of which are used by the bootloader)
- 2 KB SRAM
- 1 KB EEPROM
- Output current per port ±40 mA
- Supply voltage 7 V to 12 V
- Size: ca. 70 x 53 x 15 mm

2.5 | ArduPilot

Figure 2.5: ArduPilot, an Arduino-compatible UAV controller based on the ATmega328 (Source: SparkFun)

ArduPilot for ambitious model airplane pilots

The ArduPilot is a very interesting controller for model airplane pilots. It enables a model airplane to fly autonomously.

Tip More information can be found at:

http://diydrones.com

2.6 | LilyPad

The LilyPad was developed for e-clothes, i.e. electronically active clothing. It can be sewn directly into the clothes or the fabric. The connection to sensors and actuators takes place via ductile threads. The whole circuit can be stowed away invisibly. The LilyPad was designed by Leah Buechley and SparkFun Electronics.

Possible applications of the LilyPad

The LilyPad can be used to enhance safety by illuminating the clothes of pedestrians and cyclists in the dark.

Specifications

Figure 2.1: LilyPad Arduino (Source: Elmicro)

Specifications
- ATmega328V, older models with ATmega168V
- 16 MHz
- Programming via USB adapter (Arduino/USB)
- Supply voltage 2.7 V to 5.5 V
- 14 digital I/O pins, 6 of which are PWM-capable
- 6 analog inputs (10-bit)
- Output current per digital port 40 mA
- 32 KB flash (2 KB of which are used by the bootloader); 16 KB for ATmega168 models
- 2 KB SRAM; 1 KB for ATmega168 models
- 1 KB EEPROM; 512 B for ATmega168 models
- Size: ca. 70 x 53 x 15 mm

2.7 | USB adapter

Figure 2.6: USB adapter with FTDI chip (Source: Elmicro)

This programming adapter is available in a 3.3 V and a 5 V version.

This adapter is required for programming Arduino boards without a USB socket, e.g. the Arduino Mini. The pin layout complies with the original Arduino specifications. The adapter can also be used for communication (virtual serial interface). This is a must-have feature for your own developments! It enables you to load a *sketch* (= program) onto the board without pressing the reset button.

USB adapters are more than just technical toys

There are tons of diverse extension boards for the Arduino. When you look in the Internet, you will find new boards and useful extensions on a monthly basis. In the Arduino community, extension boards are called *shields*. They have the same footprint as the standard Arduino, so that you can simply stack them onto your Arduino (with the exception of the smaller units and the LilyPad).

Further useful boards and extensions

TURN ON YOUR CREATIVITY

FRANZIS
ARDUINO

ARDUINO
SHIELDS

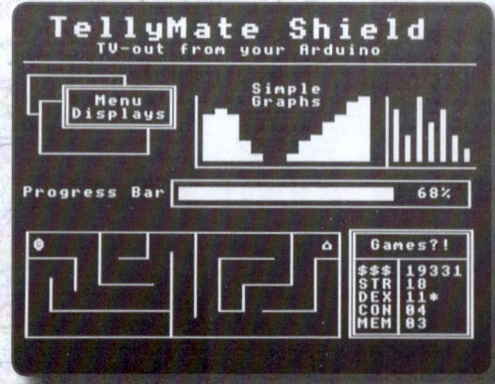

3

The various Arduino shields provide you with a multitude of creative opportunities. In this chapter, some of them are introduced along with their capabilities and options.

3.1 | Arduino ProtoShield

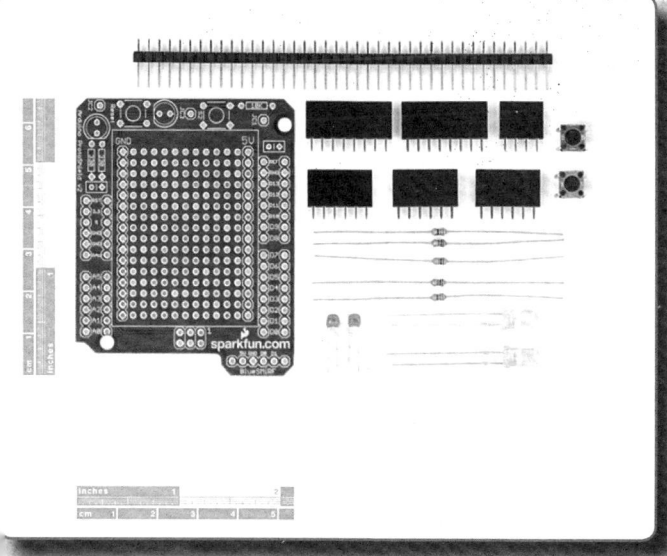

Figure 3.1: Arduino ProtoShield kit (Source: SparkFun)

When you want to develop your own projects without any soldering, the ProtoShield is a good choice. It allows for experiments on a small breadboard.

Figure 3.2: Arduino Uno with ProtoShield kit (Source: SparkFun)

3.2 | Ardumoto

Ardumoto for controlling small motors

The Ardumoto motor driver shield is the ideal choice for controlling small motors. The wires from the motors are simply connected with the screw terminals on the shield. A small program causes the motors to rotate in the desired direction and with the desired speed. Specifications comply with the integrated motor driver IC L298. The Arduino MotorShield is a similar device.

3 Arduino Shields   33

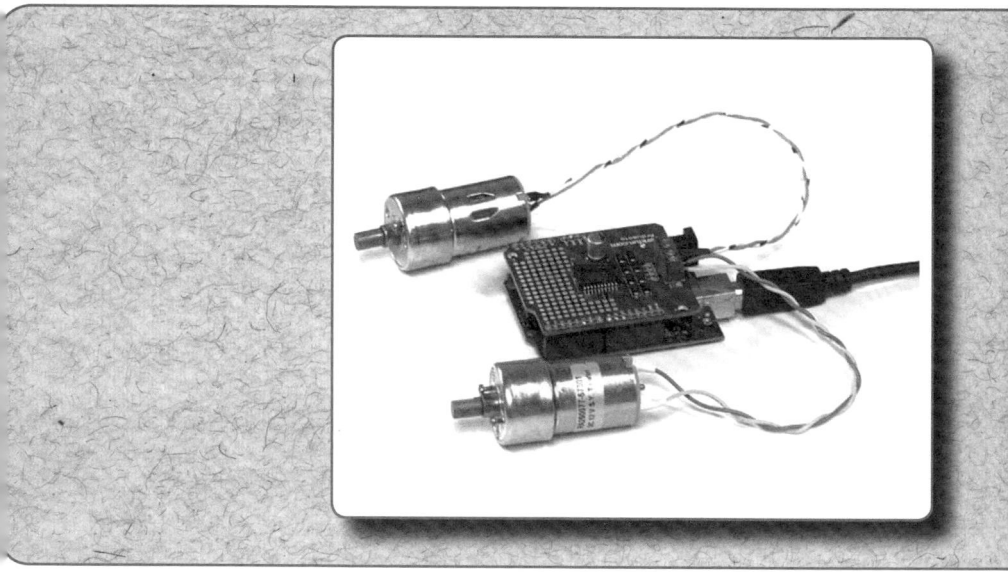

Figure 3.3: Ardumoto motor driver shield (Source: SparkFun)

3.3 | TellyMate

The TellyMate is arguably the most ingenious shield available for the Arduino. It is suitable for nearly unlimited applications and allows the presentation of data (ADC, I/Os, etc.) or just text or graphics on a TV screen, effectively converting your TV set to an Arduino display. For communicating with the TellyMate, the Arduino microcontroller uses the serial interface.

TellyMate, a very special shield

Figure 3.4: TellyMate (Source: SparkFun)

What the TellyMate can do

Features
- Arduino output on the TV set
- Composite video (PAL and NTSC)
- Stackable Arduino shield
- Works with `Serial.println()` etc.
- 38 x 25 characters
- Character representation in black and white
- Simple graphics
- Simple programming

3 Arduino Shields | 35

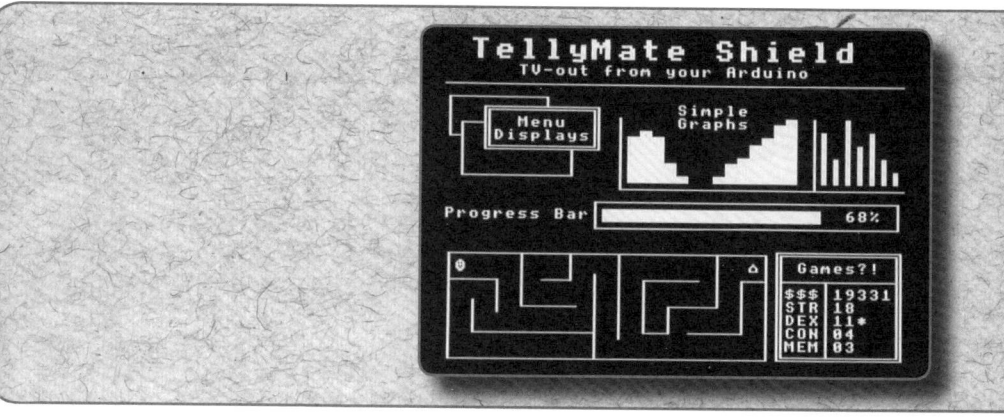

Figure 3.5: TellyMate in action (Source: SparkFun)

3.4 | XBee radio frequency modules

Figure 3.6: XBee ZNet 2.5 OEM module (Source: SparkFun)

When you want to transfer data over the air, you should consider buying XBee modules that create a wireless serial (UART) connection. With these devices, you can establish radio communication between two Arduinos or one Arduino and a PC.

Specifications

- Operating voltage: 2.8 V to 3.4 V
- Frequency: ZigBee standard, 2.4 GHz ISM band
- Transmission power: 0 dbm (1 mW)
- Sensitivity: -92 dbm
- Range: 30 m inside / 100 m outside (depending on environmental conditions)
- Input current: TX 45 mA, RX 50 mA, stand-by 10 µA
- Data rate (wireless transmission): 250,000 bps
- Data rate (interface):1200–115,200 bps
- Serial interface: 0 V/3.3 V; a 3.3 V level converter (MAX32329 is mandatory for connecting to a PC)
- Standard: compatible with ZigBee/802.15.4
- Topologies: Point-to-point, point-to-multipoint
- Size: 24.38 x 27.61 x 4 mm, 2 mm grid

3.5 | Ethernet shield

Figure 3.7: Ethernet shield (Source: Solarbotics)

The Arduino Ethernet allows the connection of an Arduino with a network or the Internet. It is based on the Wiznet W5100 Ethernet chip with a network (IP) stack that supports TCP and UDP. Furthermore, it allows for up to four simultaneous socket connections. The Arduino project provides a comprehensive library and several example programs to facilitate your first steps in the world of networking.

The Arduino Ethernet allows the connection of an Arduino with a network or the Internet

Tip You can find an overview of all available Arduino boards on:

http://arduino.cc/en/Main/Hardware

TURN ON YOUR CREATIVITY

FRANZIS
ARDUINO

COMPONENTS IN THE TUTORIAL KIT

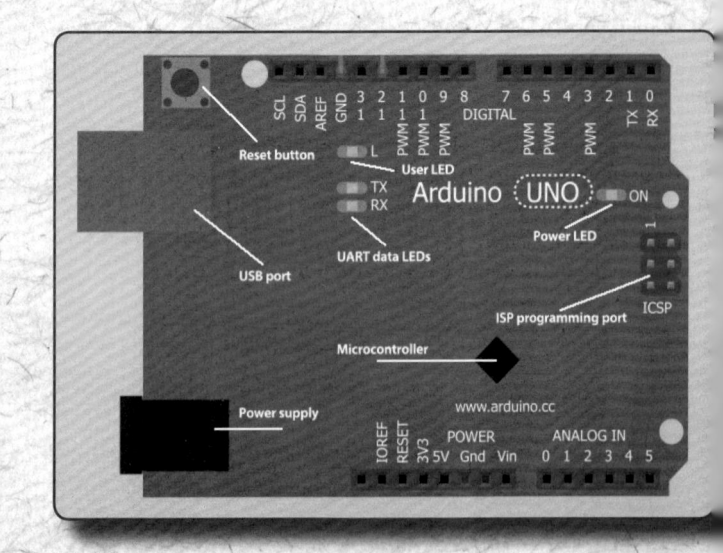

Reset button
User LED
TX
RX
UART data LEDs
USB port
Power supply
Microcontroller
Power LED
ON
ISP programming port
ICSP

SCL SDA AREF GND 3 2 1 0 9 8 7 6 5 4 3 2 1 0
1 1 1 1 DIGITAL TX RX
PWM PWM PWM PWM PWM PWM

Arduino UNO

www.arduino.cc

IOREF RESET 3V3 5V Gnd Vin 0 1 2 3 4 5
POWER ANALOG IN

4

In the previous chapters, you were introduced to the Arduino and its uses and you learnt a bit about programming. Now it is time to proceed with practical experiments with your Arduino Uno. First of all, we take a look at the components in the tutorial kit.

4.1 | A survey of the components

All components in the tutorial kit

- 1 Arduino Uno
- 1 breadboard
- 2 push-buttons
- 1 NPN transistor BC548C
- 1 silicon diode 1N4148
- 1 piezo buzzer
- 1 red LED
- 1 green LED
- 2 yellow LEDs
- 3 resistors 1.5 kΩ
- 1 resistor 4.7 kΩ
- 1 resistor 47 kΩ
- 1 resistor 10 kΩ
- 1 resistor 68 kΩ
- 1 trim potentiometer 10 kΩ PT10
- 1 capacitor 1 μF
- 1 insulated jump wire ca. 1 m

4.2 | Arduino Uno

The microcontroller is controlled via the USB port

First, we take a look at the microcontroller board. This board comprises a USB port of type B for the physical connection between the Arduino and a PC. This connection is used to transfer *sketches* (that is, programs) from the PC to the Arduino as well as data in the other direction. On the Arduino board, a controller serves as communication interface and converts the USB signals of the PC to serial signals the Arduino microcontroller can understand. After installation, the Arduino board is shown as a virtual COM port in the device manager.

Tip More information about the serial interface can be found at:

http://en.wikipedia.org/wiki/Serial_port

The Arduino board with its hardware serves as the basis for the experiments described in this tutorial kit. For the larger experiments, you need the little *breadboard* which allows you to attach other components and connect them with the Arduino.

4 Components in the Tutorial Kit | 41

4.3 | Ports and LEDs of the Arduino Uno

All of the Arduino ports are accessible via pin headers. The LED labeled ON shows that the microcontroller board is powered on, no matter whether the USB connection or an external power supply is used. The LED labeled L is hard-wired to the Arduino digital pin 13 and serves as user LED. It can be used to monitor the activity of a program. The pin headers around the board can accept both wires and directly connected components.

LED L and its function

The LEDs TX and RX indicate data traffic on the serial interface. For instance, they blink alternately when a program is transferred, and notify you about communication processes.

LEDs TX and RX and their function

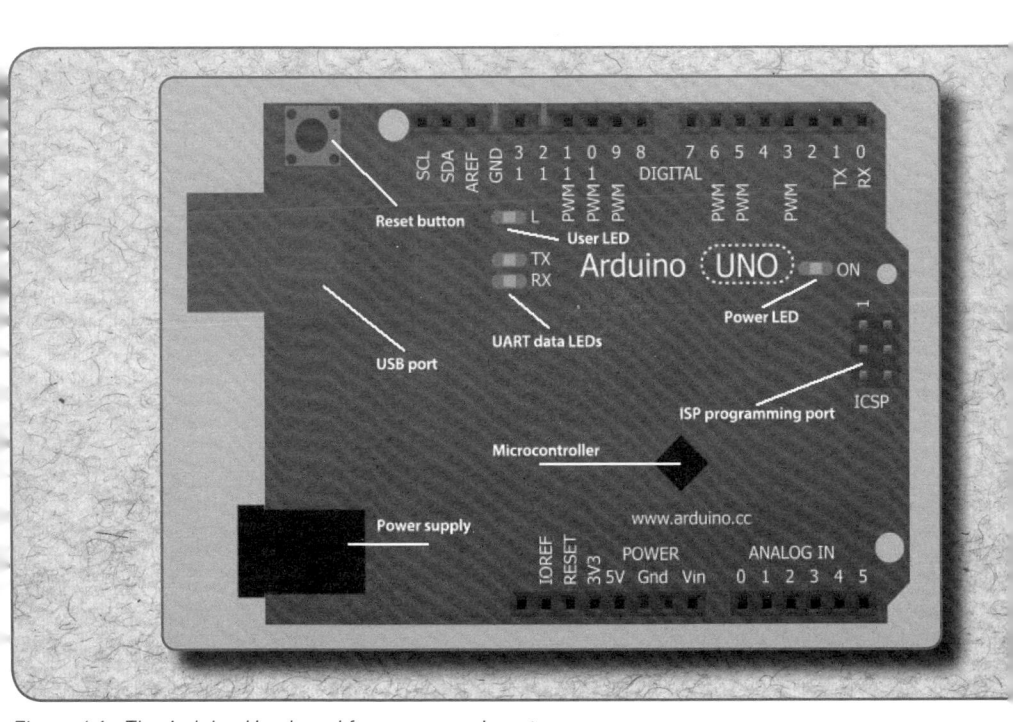

Figure 4.1: The Arduino Uno board for your experiments

4.3.1 | Quick reference guide: the board

- Reset button (To reset the controller to the initial state.)
- USB port (Establishes the connection to the PC; is used for programming and for the transfer of data between the user program (sketch) and the PC.)
- Power supply (With larger current consumers or weak USB ports, the board can be powered with an external device via this port; supply voltage between 7 V and 12 V.)
- LED L (User LED, hard-wired with pin 13; can be used freely, e.g. as a »program run« LED.)
- LED RX/TX (Shows communication on the serial interface.)
- Microcontroller (The heart of the board; includes preprogrammed bootloader that ensures the transfer of programs via USB.)
- ICSP port (Only necessary when you want to transfer programs by means of an Atmel ISP programmer, e.g. to reprogram the bootloader or when you do not want to work in the Arduino IDE.)

Tip More information and updates on this board can be found at:

http://arduino.cc/en/Main/ArduinoBoardUno

4.3.2 | Quick reference guide: the ports

The most important ports

- IOREF (Used with shields in order to adapt the board voltage to the shield voltage.)
- RESET (The reset pin of the controller is lead through this port. When connecting this pin with GND, the board is reset.)
- 3V3 (Here you can use a voltage of 3.3 V.)
- 5V (Here you can use a voltage of 5 V.)
- GND (Ground.)
- Vin (For the external power supply of the Arduino; cf. data sheet for the Arduino at *www.arduino.cc*.)
- ANALOG IN 0 to 5 (Analog measurement inputs for voltages between 0 V and 5 V; »voltage metering«.)
- RX 0 (Receiver input of the serial interface and digital port.)
- TX 1 (Sender output of the serial interface and digital port.)
- 2 (Digital port.)
- PWM 3 (PWM output and digital port.)
- 4 (Digital port.)
- PWM 5 (PWM output and digital port.)
- PWM 6 (PWM output and digital port.)
- 7 (Digital port.)
- PWM 9 (PWM output and digital port.)
- PWM 10 (PWM output and digital port.)
- PWM 11 (PWM output and digital port.)
- 12 (Digital port.)
- 13 (Digital port, connected to LED L)
- AREF (External reference voltage input for the ADC.)
- SDA (Data port of the I^{2}C bus [»TWI«].)
- SCL (Clock port of the I^{2}C bus [»TWI«].)

4.4 | Power supply

Alternatives to the USB power supply

To supply your Arduino with power, you can use the USB connection or a DC mains adapter (ca. 9 V/500 mA DC). It is not recommended to supply power via USB when larger current consumers are attached to the microcontroller board, as this can affect the USB port.

4.5 | Reset button

The reset button causes a restart of the program. This has the same effect as cutting off the power supply of the board and then reconnecting it.

4.6 | ISP port

Usage of the ISP port

The ISP port is used to program the microcontroller by means of an Atmel ISP Programmer and to upload the bootloader. You will not need this port for your experiments. The bootloader is pre-installed.

Tip Connect your Arduino board via a USB hub. If you accidently cause a short-circuit in your experiments, you will kill only the USB hub and not the USB port of your PC.

4.7 | Safety notes

The Arduino board is largely protected against mistakes. It is nearly impossible to damage your PC. However, the connections of the USB port on the bottom plate of the board are not insulated. Putting the board onto a metal conductor may cause a higher current, which could damage the PC and the board.

Be sure to follow these safety procedures!

Safety first:
How to protect your Arduino board

- Avoid any contact between metal objects and the bottom plate of the board, or insulate the whole bottom plate by means of a nonconductive protective plate or insulation tape.
- Keep away any power supplies, other power sources, and live conductors with more than 5 V from the board.
- If possible, do not connect the board directly to the PC; do so via a hub. Most of these devices come with an effective protective circuit. Even if anything happens, only the hub will be damaged, but not the PC.

TURN ON YOUR CREATIVITY
FRANZIS
ARDUINO

USE OF THE
COMPONENTS

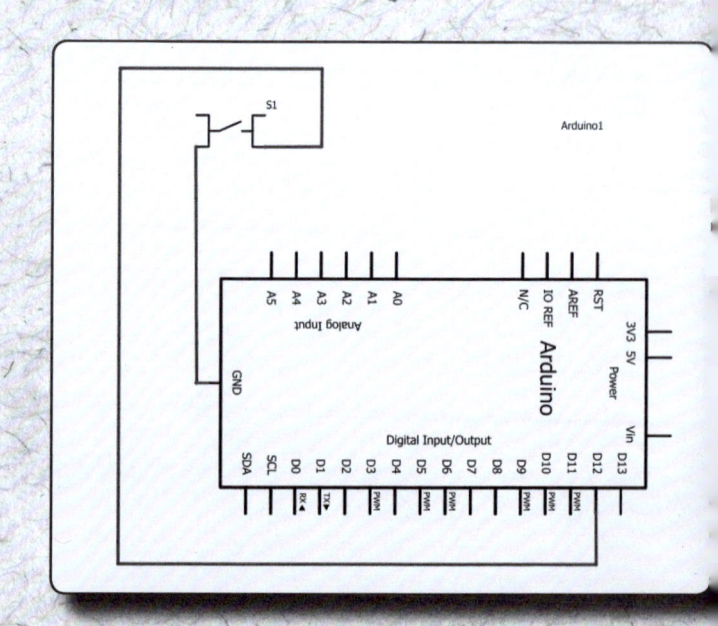

5

In this chapter, we will introduce the most important components of the tutorial kit and their use. However, only hands-on experimentation can give you the necessary experience.

The components, their use, and the first steps of experimentation

5.1 | Jump wire

You will find some jump wire in the tutorial kit. It is used for connections with the board. Cut it into pieces of suitable length (as described in the following experiments) with wire cutters or scissors, and remove the insulation on both ends. Keep the wires! You will need them quite often.

Prepare the jump wire for your experiments

Figure 5.1: *Wire wrap and prepared wire strap*

5.2 | Breadboard

The breadboard is prepared for the experiments

You can build the circuits on the breadboard without soldering. The contacts are connected from A to E and from F to J (see Fig. 5.2). The breadboard consists of 20 columns and 10 rows (A to J).

It is helpful to file the connection wires a little, so that the wire ends form a sort of wedge. This way, the components can be inserted into the breadboard more easily. In case you have any difficulties with inserting components into the breadboard, your best option is to use precision flat-nose pliers.

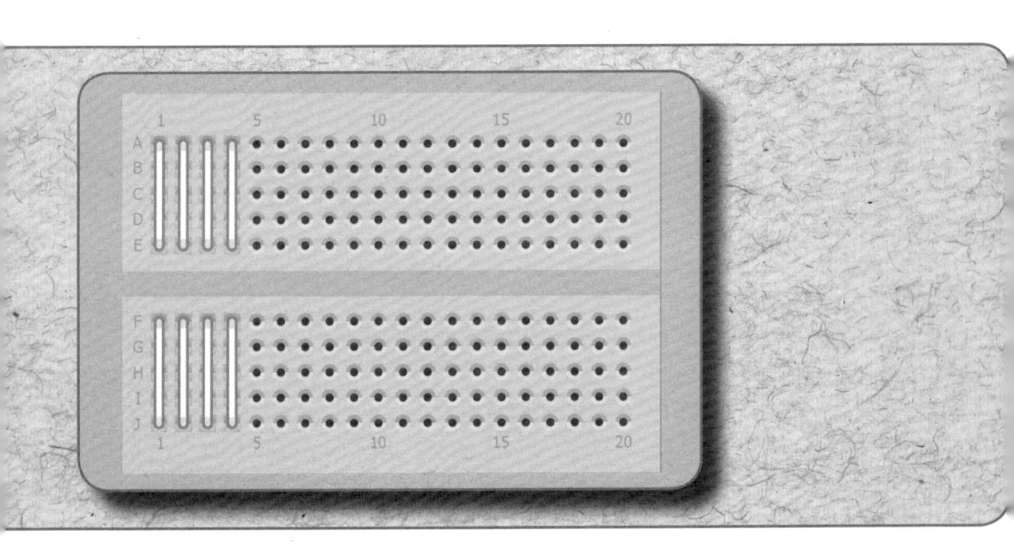

Figure 5.2: The tiny breadboard

5.3 | Push-buttons

A push-button is very similar to a switch: It closes or disconnects an electric circuit. The difference, however, is that the push-button does not remain in its position, but returns to the original position, when you release it. The push-buttons in the tutorial kit come with four pins, two of which are always connected, as you can see in the illustration.

The push-button and its use

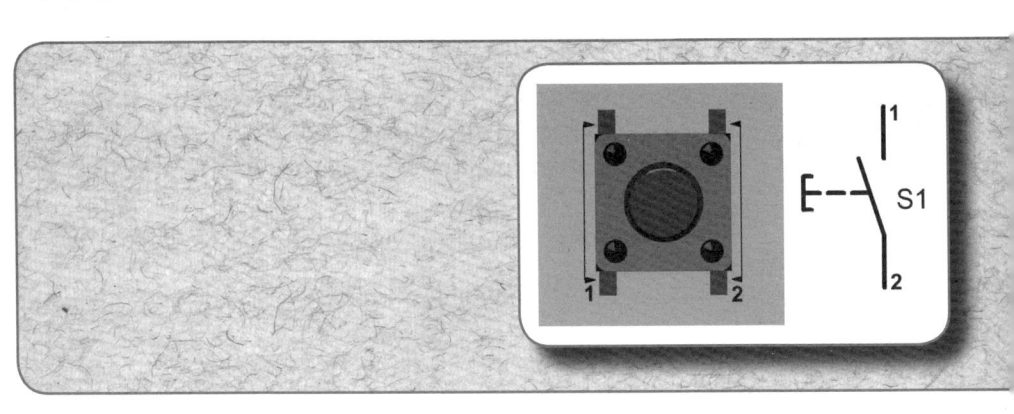

Figure 5.3: The push-button

5.4 | Resistors

Resistors are used to limit the current, to adjust the operating point, and as voltage dividers. The unit of measurement for electrical resistance is the Ohm (Ω, the Greek letter omega). With the use of prefixes like k (kilo, 1000) and M (mega, 1 million), you can write large values in abbreviated form:

1 kΩ = 1000 Ω
10 kΩ = 10.000 Ω
100 kΩ = 100.000 Ω
1 MΩ = 1.000.000 Ω
10 MΩ = 10.000.000 Ω

What does the colour code on resistors mean?

In circuit diagrams, the Ω sign is often omitted, and 1 kΩ is often abbreviated as 1k. The resistance value is colour-coded on the resistor, usually in form of three coloured rings with a fourth one for the precision at a little distance.

Metal film resistors have a tolerance of only 1%. This is indicated by a brown tolerance ring, which is slightly broader than the other four coloured rings. This is to avoid any confusion with the normal value ring »1«.

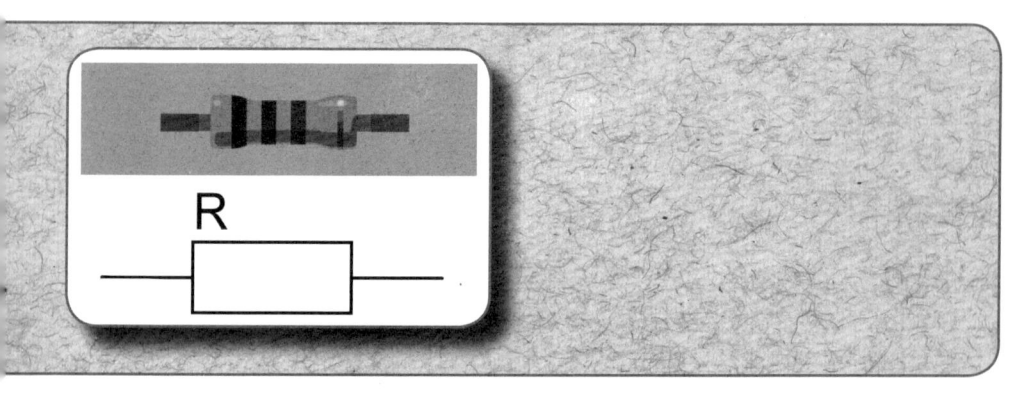

Figure 5.4: A resistor and the circuit diagram symbol

There are also resistors with a tolerance of ±5% and the values of the E24 series, where every decade contains 24 values with nearly identical difference to the neighboring values.

5 Use of the Components

The resistors of the E24 standard series are as follows:

1.0 / 1.1 / 1.2 / 1.3 / 1.5 / 1.6 / 1.8 / 2.0 / 2.2 / 2.4 / 2.7 /
3.0 / 3.3 / 3.6 / 3.9 / 4.3 / 4.7 / 5.1 / 5.6 / 6.2 / 6.8 / 7.5 / 8.2 / 9.1

The colour code is read beginning with the ring nearest to the end of the component. The first two rings represent the digits of the resistance value, the third one the multiplicator.

How to read the colour code on resistors

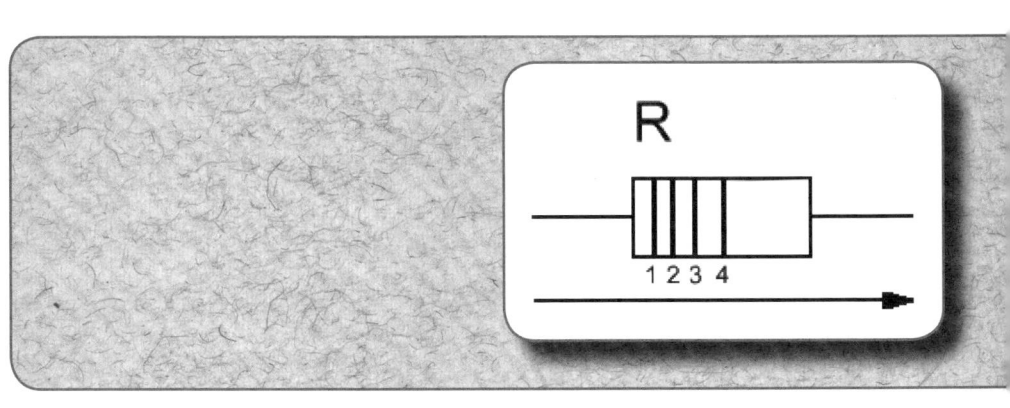

Figure 5.5: Reading the resistance value

A resistor with a yellow, a purple, a brown and a gold ring has a resistance of 470 Ω and a tolerance of ±5%. Now try to identify the resistance value of the resistors in the tutorial kit.

In addition to fixed resistors, there are also variable resistors called potentiometers. The resistance value is nearly always imprinted on the component in numerical form. This value means the maximum adjustable value. The potentiometer comes with three pins: two at the ends of the resistive element and one for the wiper. At the ends of the

Variable resistors: potentiometers and trimmers

circular resistive element, you get a fixed resistance value. With the wiper, you can tap part of this resistance. You can use a potentiometer for instance to create a variable resistance divider. For this purpose, you apply the supply voltage to the ends of the potentiometer and tap the voltage difference to ground at the adjustable middle pins.

Colour	Ring 1	Ring 2	Ring 3 (Factor)	Ring 4 (Tolerance)
Silver	—	—	$1 \times 10^{-2} = 0.01\ \Omega$	±10%
Gold	—	—	$1 \times 10^{-1} = 0.1\ \Omega$	±5%
Black	0	0	$1 \times 10^{0} = 1\ \Omega$	
Brown	1	1	$1 \times 10^{1} = 10\ \Omega$	±1%
Red	2	2	$1 \times 10^{2} = 100\ \Omega$	±2%
Orange	3	3	$1 \times 10^{3} = 1\ k\Omega$	
Yellow	4	4	$1 \times 10^{4} = 10\ k\Omega$	
Green	5	5	$1 \times 10^{5} = 100\ k\Omega$	±0.5%
Blue	6	6	$1 \times 10^{6} = 1\ M\Omega$	±0.25%
Purple	7	7	$1 \times 10^{7} = 10\ M\Omega$	±0.1%
Gray	8	8	$1 \times 10^{8} = 100\ M\Omega$	
White	9	9	$1 \times 10^{9} = 1000\ M\Omega$	

Figure 5.6: Meaning of the colours on resistors with four coloured rings

5 Use of the Components | 53

Instead of a solid control shaft and a sturdy housing, trimmers have a slotted disk. Furthermore, they are smaller and less robust. The wiper is adjusted with a screwdriver. A potentiometer with control shaft is usually installed in a device housing and provided with an adjusting knob. The trimmer P1 on our board is an adjustable resistor connected as a variable voltage divider.

How to identify a trimmer

Figure 5.7: A trimmer

Tip Using the search term »resistance calculator«, you can find many websites for converting colour codes to resistance values.

5.5 | Capacitors

Constituent parts of a capacitor

Basically, a capacitor consists of two metal surfaces facing each other and an insulating layer between them. When you apply voltage, an electrical field is generated between the capacitor plates, and energy is stored in this field. The larger the surface of the capacitor plates, the more energy can be stored.

The capacitance

The capacitance of a capacitor is specified in Farad (F). The capacitors generally used in electronics and the ones in this tutorial kit have a capacitance between 10 nF (0,00000001 F) and 1000 µF (0,001 F). The insulating material (dielectric) provides for a higher capacitance compared to air insulation. Ceramic disk capacitors come with a special insulating material to achieve high capacities with a small footprint.

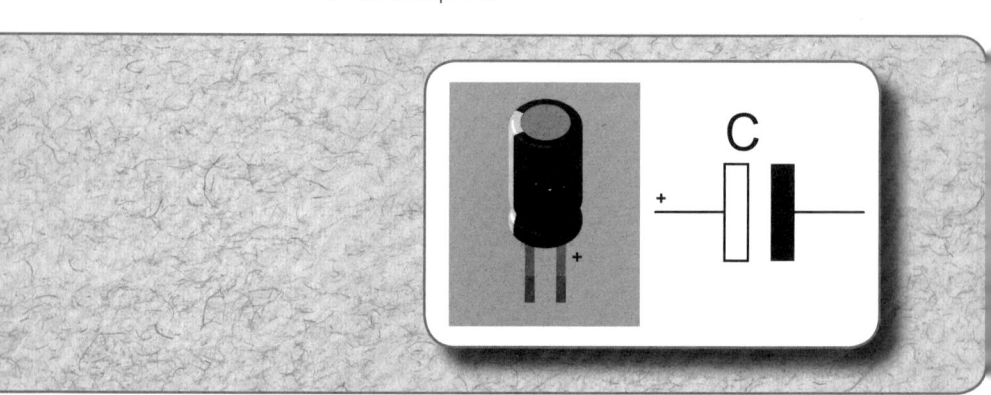

Figure 5.8: A capacitor and the circuit diagram symbol

5 Use of the Components

The figure shows an electrolytic capacitor like the ones in the tutorial kit. With this type of capacitor, you have to pay attention to the polarity, because it will explode when inserted the wrong way! In the circuit diagram, the negative terminal is represented by a solid bar, and the positive terminal by an empty rectangle. On the component itself, the negative terminal is marked by a white bar. Another distinguishing feature is the different length of the pins. As with an LED, the longer leg is the positive terminal, the shorter one the negative.

Check the polarity of a capacitor!

Non-polarised capacitors do not have any terminal markings. In circuit diagrams, they are represented by two parallel (black) bars without terminal designation.

How to identify non-polarised capacitors

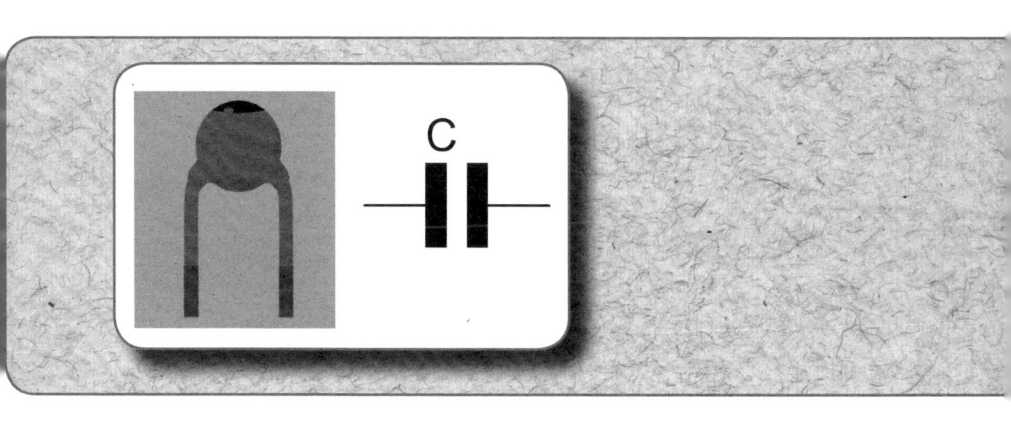

Figure 5.9: Circuit diagram symbol of a ceramic capacitor

Info You do not need any ceremaic capacitors for our experiments. This description is added only for the sake of completeness.

5.6 | Piezo buzzer

Working principle of a piezo buzzer

The piezo buzzer acts as a small speaker, sensor or microphone. It is built like a ceramic disk capacitor, but with an additional bias voltage applied to the dielectric. This way, mechanical tension and voltage are coupled. The piezoelectric effect also occurs in similar form in natural quartz crystals. A good example for such a piezo-electric effect is an electrical lighter. However, the voltage of such a device is much higher than that of our piezo buzzer.

Figure 5.10: A piezo buzzer and its circuit diagram symbol

5.7 | LEDs

The tutorial kit contains red, yellow and green light emitting diodes (LEDs). LEDs belong to the category of optoelectronics. The main difference to regular diodes (rectifier diodes etc.) is that they light up when electric current runs through them. The avalanche voltage varies with type and colour. The anode of an LED is marked by the longer pin or, when you look inside the diode, by the small terminal next to the cup. The cathode pin (ground) is flattened at the housing. When dealing with LEDs, always use a series resistor to limit the operating current so that the diode does not blow.

The LEDs in the tutorial kit

Caution Always pay attention to the polarity of the LED. Never connect the diode directly to a battery or a USB power supply. This will destroy the diode (and possibly the Arduino board, too).

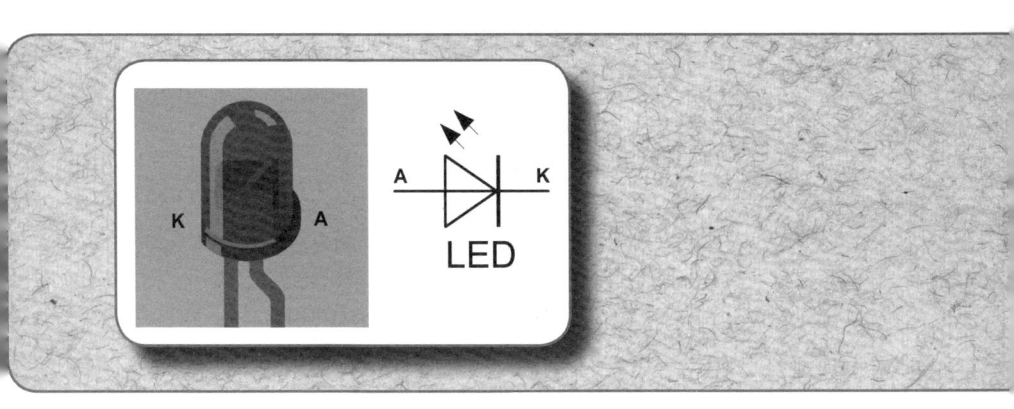

Figure 5.11: An LED and its circuit diagram symbol

LEDs vary in colour, current consumption, diameter and brightness. The luminosity is always specified in candela, in most cases in millicandela (mcd). The tutorial kit contains 5 mm low current LEDs that exhibit their maximum luminosity already at 2 mA. Because of the low current consumption, these LEDs are ideally suited for CMOS and microcontroller applications. The following examples show how to calculate the resistance value of a series resistor for an LED:

Units of measurements for LEDs

The avalanche voltage amount to the following values:
- Red LEDs: typically 1.6 V, max. 2 V
- Orange LEDs: typically 2.2 V, max. 3 V
- Green LEDs: typically 2.7 V, max. 3.2 V
- Yellow LEDs: typically 3.4 V, max. 3.2 V

How to calculate the resistance of a series resistor

Example calculation for the series resistor of a red low current LED (2 ma):

```
Resistance value of the series resistor
= (supply voltage - LED voltage)/LED current
= (5 V - 2 V)/0.002 mA
= 1500 Ω = 1.5 kΩ
```

5.8 | Diode

Usage of diodes

A diode is an electrical component. You can think of it as a sort of electronic valve that allows current only to flow in one direction. Diodes are distinguished according to the basic material: germanium (Ge) or silicon (Si). The experiments in this kit are conducted with silicon diodes of the type 1N4148. These popular Si-diodes can tolerate currents up to 100 mA. As with LEDs and capacitors, you have to pay attention to the polarity. The negative terminal is marked by a small ring at the edge of the housing.

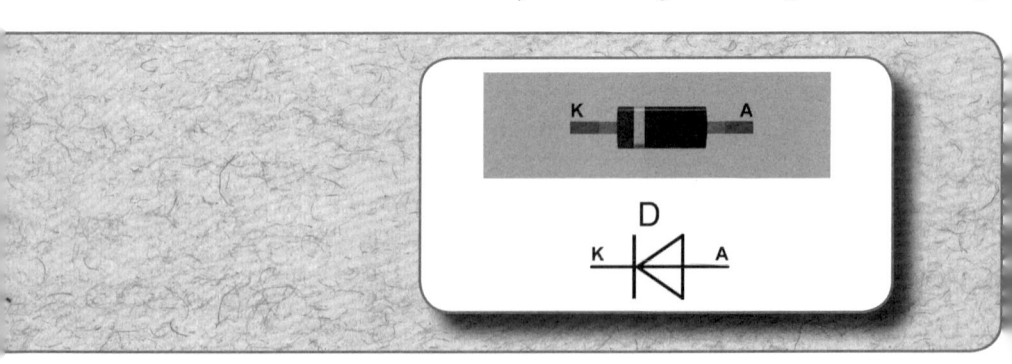

Figure 5.12: A diode and its circuit diagram symbol

5.9 | Transistors

Transistors are components for boosting small currents. A low base current causes a high collector current. These components have three pins: base (B), collector (C), and emitter (E).

The transistor and its components

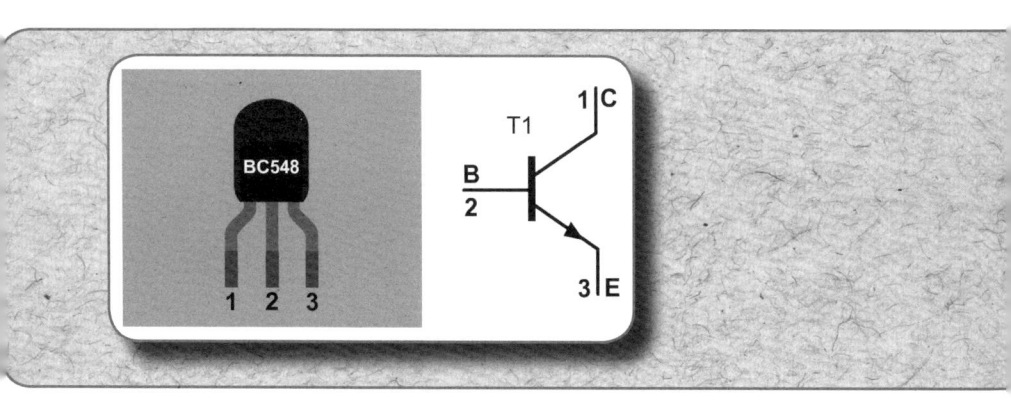

Figure 5.13: An NPN transistor and its circuit diagram symbol

In the experiments, the transistor BC548C is used. It is a universal small signal transistor for low voltages and currents.

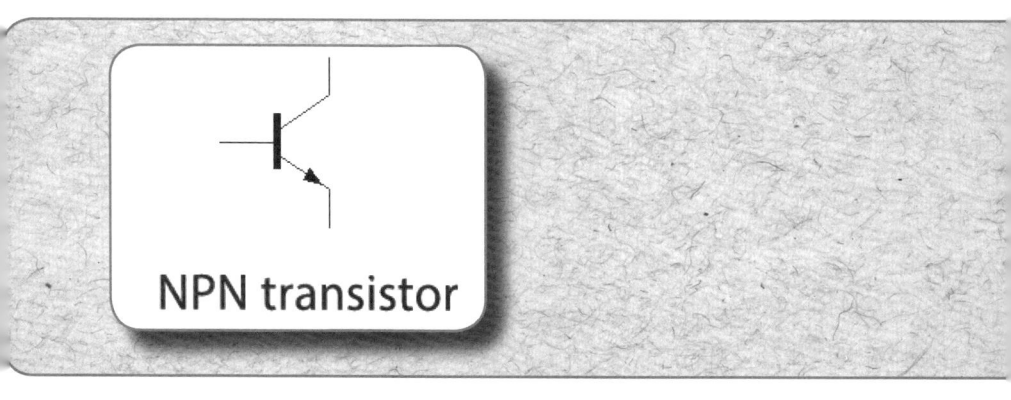

Figure 5.14: An NPN transistor

PNP transistor

Figure 5.15: A PNP transistor

The PNP transistor works like the NPN transistor, but with reversed polarity. This time, the emitter is the common positive terminal.

NPN and PNP transistors

The difference between these two types of transistors is indicated by the emitter arrow in the circuit diagram symbol: In the symbol for an NPN transistor, the arrow for the direction of current points away from the base whereas in the symbol for a PNP type, it points towards the base.

For functional tests of transistors, an off-the-shelf multimeter will suffice. Set the multimeter to the Ohm range Fix the test tip of the ohmmeter at the base and touch consecutively collector and emitter with the other tip Depending on polarity, the ohmmeter indicates »pass« or »infinite« at both terminals. Now reverse the polarity of the ohmmeter and repeat the process. The transistor is OK if it now exhibits the exact opposite behaviour.

5 Use of the Components

The base-emitter voltage must not exceed the indicated maximum value. Higher voltages can cause damage to the transistor. In each and every circuit, the base voltage is limited by the appropriate base resistors. Most modern multimeters include a diode testing feature that you can use to check this. The equivalent circuit diagrams show the alignment of the diodes. If not otherwise indicated, the pin assignment of transistors is always depicted as if you were looking at it from below.

Check the maximum values for the base-emitter voltage

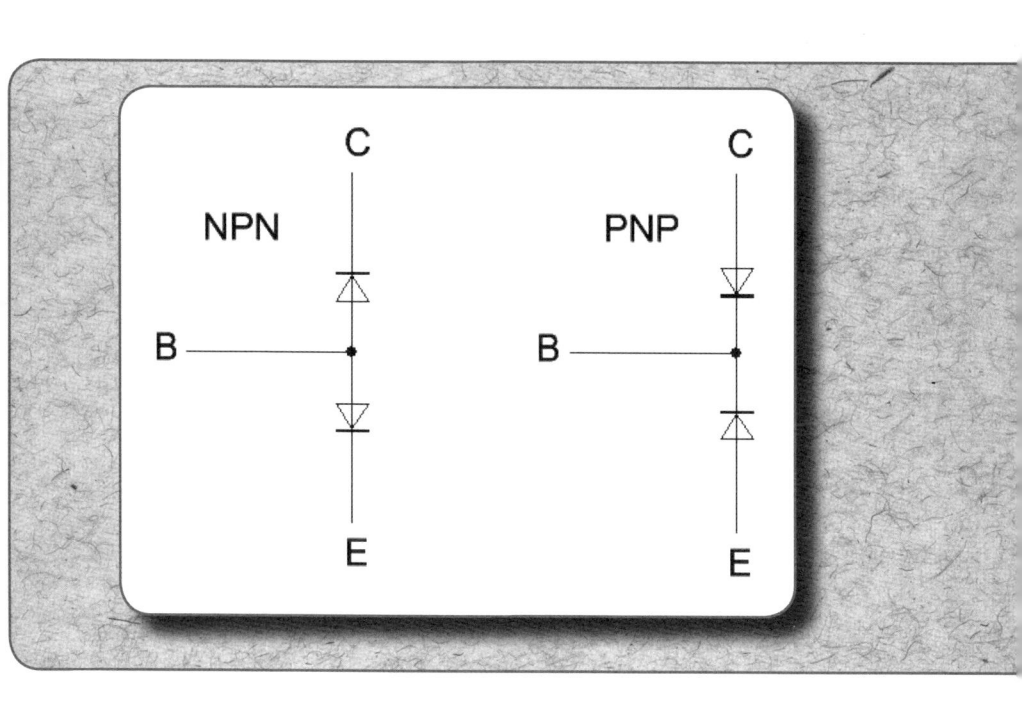

Figure 5.16: You can think of a transistor as a circuit made of diodes connected in the way this figure shows. This makes it easier to test a transistor for defects.

TURN ON YOUR CREATIVITY

FRANZIS
ARDUINO

INSTALLATION OF THE PROGRAMMING ENVIRONMENT

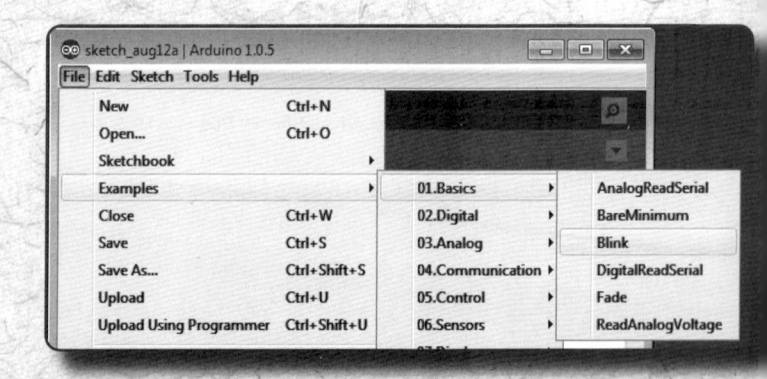

6

Before you can begin with experimenting and programming, you have to take some preparatory steps. These include installing the drivers for the virtual COM port (serial interface) on the PC and installing the programming/development environment for the Arduino.

Some final preparation steps before you can begin

6.1 | Installation on Windows

As a first step, download the newest version of the Arduino programming environment from *www.arduino.cc*. The Arduino is continually developed further, and every few weeks a new programming environment (Integrated Development Environment, IDE) with improvements and new features is available. Hence a glance at the website is always worthwile.

The Arduino IDE is continually improved – visiting the website is always worthwile!

Unzip the archive file (e. g. by using 7.zip, *www.7-zip.de*) and copy the folder to your favored location on the hard disk. A proper installation is not necessary. You can start *Arduino.exe* immediately. But for now, we will postpone this action.

Tip It is not mandatory to use system drive *C:* as installation location. In case Windows ceases to function and you have to do a reinstallation, your Arduino programs may vanish when you place them on the system drive.

A step-by-step guide to installation

Next, you connect the Arduino Uno with your PC. Windows then tries to install a new driver. However, the initial installation will fail, because Windows does not know where the drivers are located on your computer.

Open the Device Manager in the Windows Control Panel.

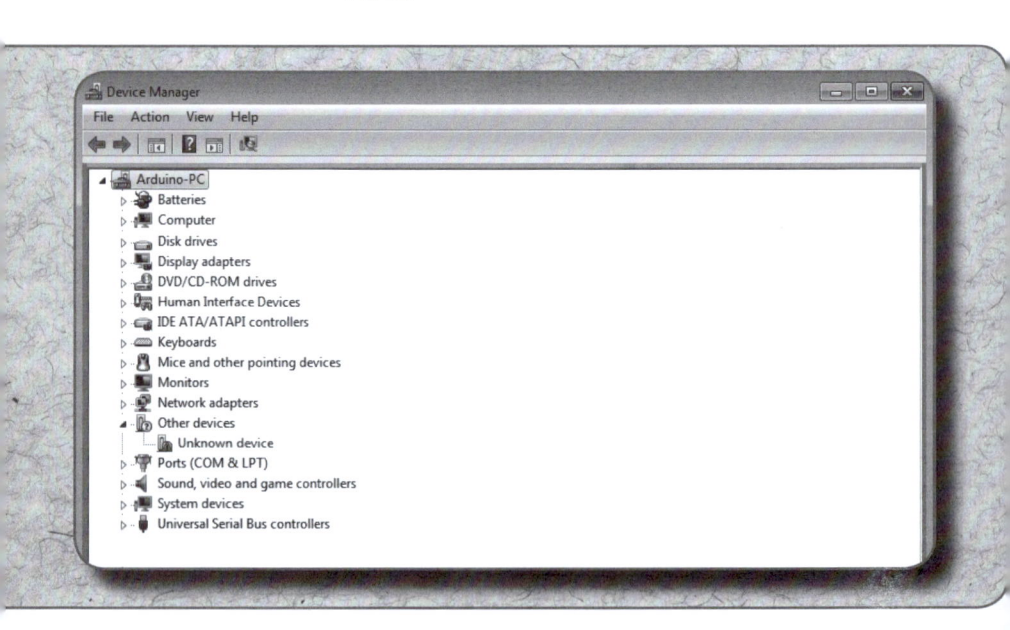

Figure 6.1: After attaching the Arduino to the PC, the Device Manager shows the icon »unknown device«

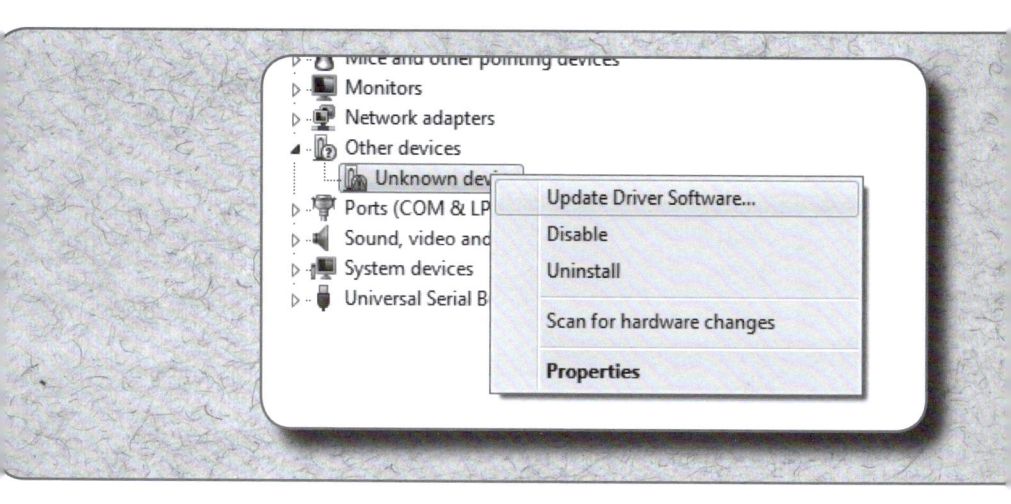

6 Installation of the Programming Environment 65

Figure 6.2: Right-click to open the context menu and choose **Update driver software**

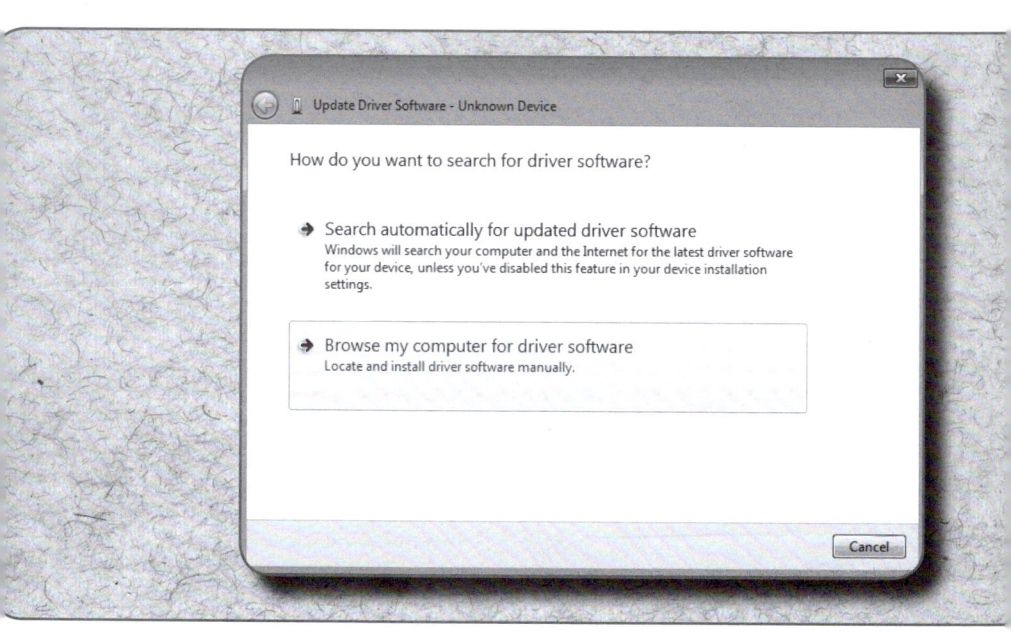

Figure 6.3: Choose **Browse for driver software on your computer**

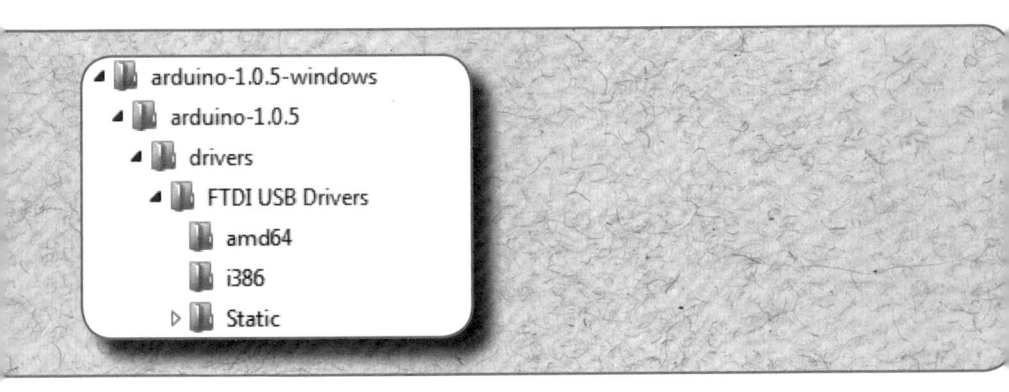

```
▲ 📁 arduino-1.0.5-windows
    ▲ 📁 arduino-1.0.5
        ▲ 📁 drivers
            ▲ 📁 FTDI USB Drivers
                📁 amd64
                📁 i386
                ▷ 📁 Static
```

*Figure 6.4: Select the path to the driver. The drivers can be found in the folder **Arduino** and the subfolder **drivers**. Select the driver folder and confirm with **OK**.*

Windows Security

Would you like to install this device software?

Name: Arduino USB Driver
Publisher: Arduino LLC

☐ Always trust software from "Arduino LLC".   [Install]   [Don't Install]

ⓘ You should only install driver software from publishers you trust. How can I decide which device software is safe to install?

Figure 6.5: Since the driver is unsigned (signing it would be a very complex and expensive effort), Windows warns against installing it. It is safe to install the driver anyway.

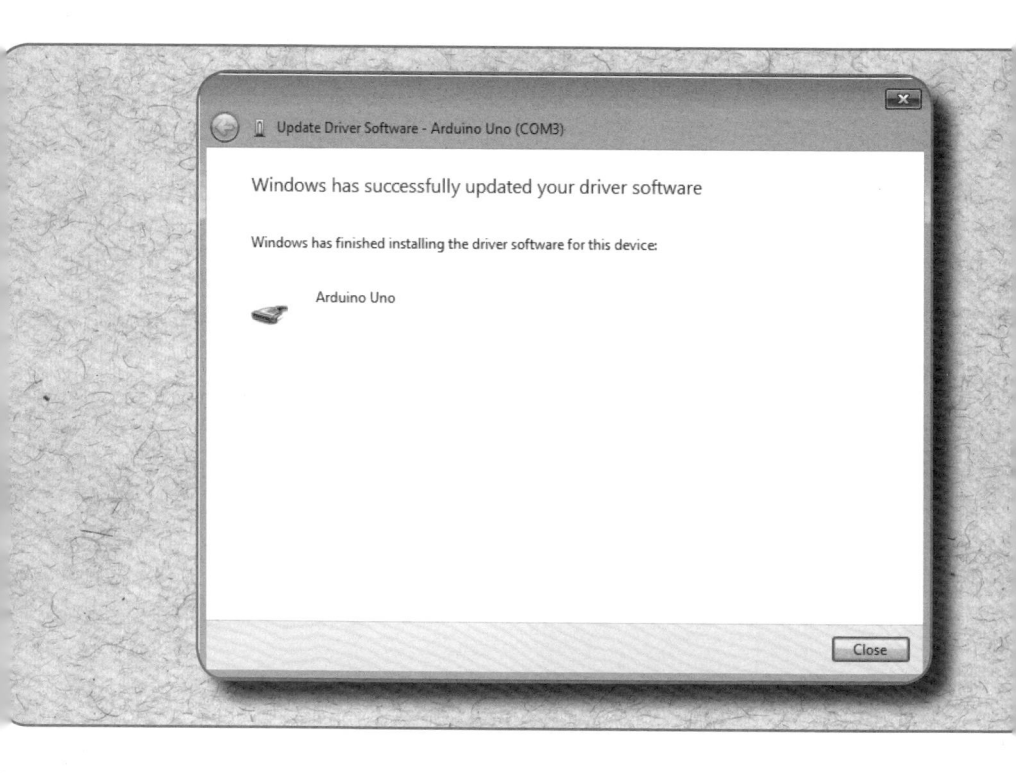

6   Installation of the Programming Environment   67

Figure 6.6: After successful installation of the driver, you can immediately use the Arduino!

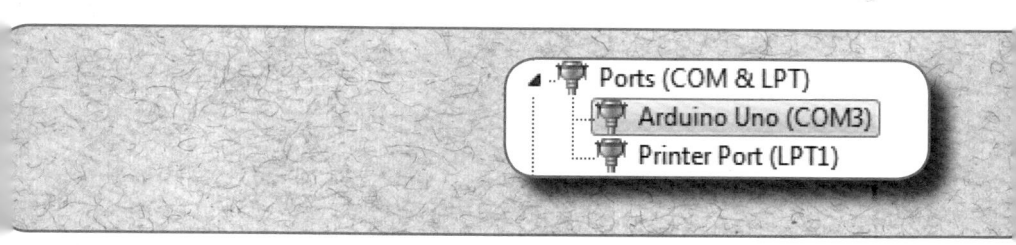

*Figure 6.7: The Arduino Uno is shown in the **Ports** section of the Device Manager. The entry contains the revision number (R3 or higher) and the number of the COM port.*

Open the development environment by double-clicking *Arduino.exe*. It is a good idea to create a shortcut on the desktop, because you will use this environment quite often.

Speeding up: How to create a desktop shortcut

Tip To create a desktop shortcut, simply right-click on the appropriate object and choose *Create desktop shortcut*. Now the icon is placed on your desktop, where you can start it as usual by double-clicking.

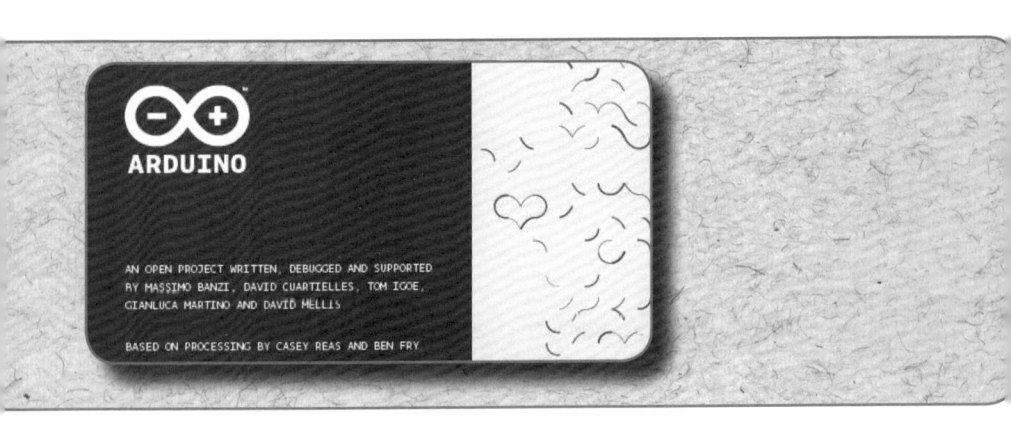

Figure 6.8: After starting the Arduino IDE, a splash screen is shown temporarily

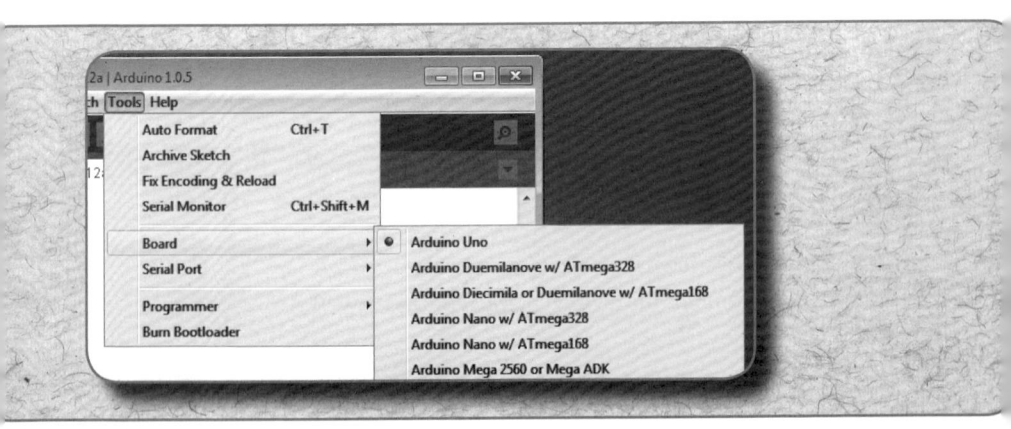

*Figure 6.9: Under **Tools\Boards**, choose the Arduino Uno*

6   Installation of the Programming Environment   69

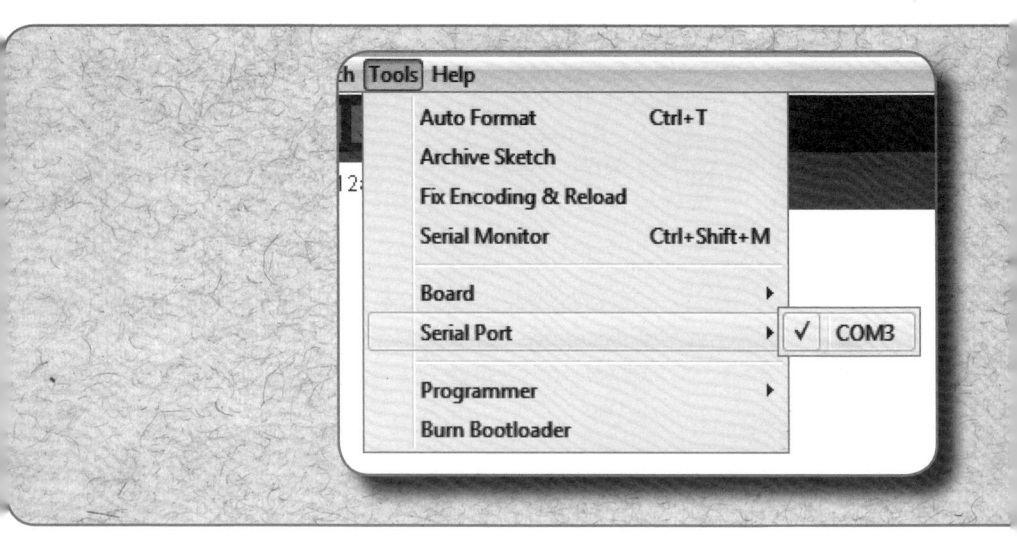

Figure 6.10: Next, choose the assigned COM port (here COM14). On your system, the port may have another number, depending on how many ports are already in use on your computer.

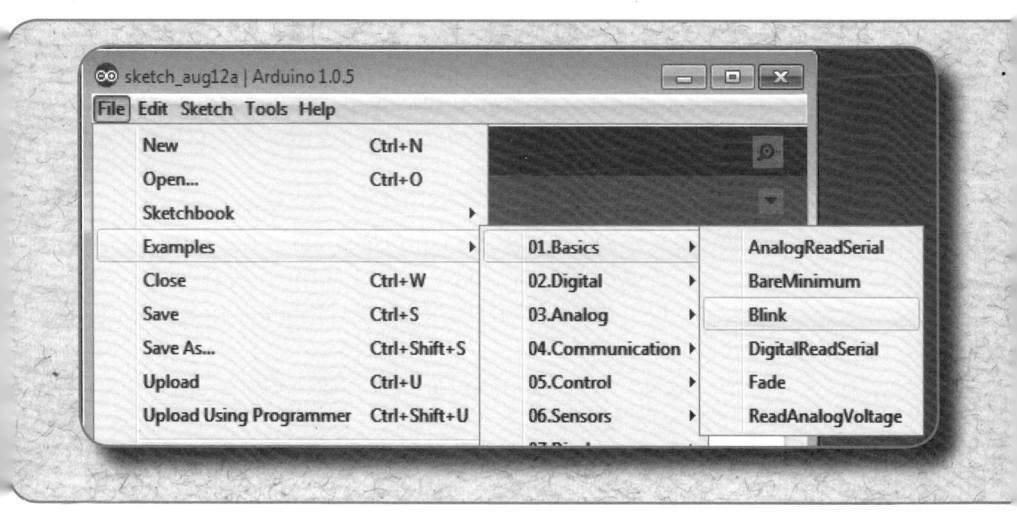

*Figure 6.11: Now test the board with a small testing program that is included with the Arduino. Open **File\Samples\Basics** and choose **Blink**.*

Figure 6.12: Upload the program to the Arduino Uno. Click the right arrow or hit `Ctrl` + `U`. *The program is now transferred to the microcontroller. You can verify the data transfer by monitoring the RX/TX LED on the board. It will blink rapidly when data transfer is in progress.*

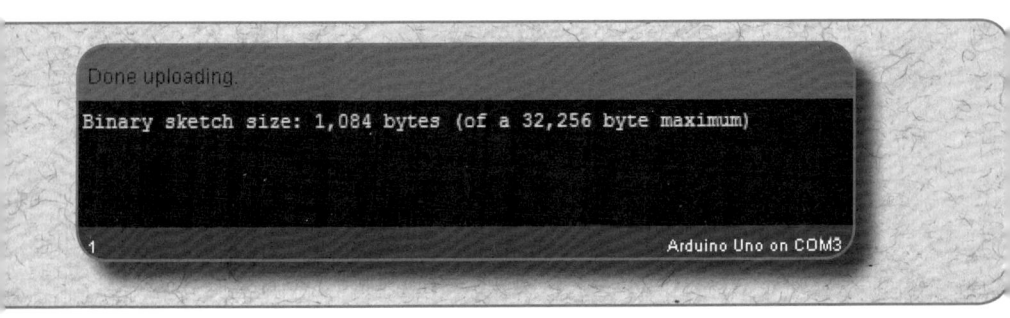

Figure 6.13: After transferring the program successfully, the programming environment shows this view. The LED L on the board will blink slowly.

Troubleshooting When the data transfer fails, check the settings fo Board and Port. Make sure that the board is shown correctly in the Device Manager and that it is properly installed.

Tip You can find further information concerning the installation here:

http://arduino.cc/en/Guide/Windows

6.2 | Installation on Mac OS X

As a first step, download the newest version of the Arduino programming environment from *www.arduino.cc*. The Arduino is continually developed further, and every few weeks a new programming environment (Integrated Development Environment, IDE) with improvements and new features is available.

How to install the Arduino IDE on Mac OS X

You get a disk image that you have to mount as usual. Then a Finder window will be opened. Install the Arduino IDE by dragging the disk image to the *Applications* folder. The image contains two installation packages for the USB drivers. Choose the appropriate package for your system architecture. On a Mac that was built before 2006, you have to use the PPC version instead of the Intel version. The installation follows the usual process on the Mac. You have to provide your password in order to continue with the installation because the driver executes some kernel extensions.

After completing the installation, you can find the Arduino programming environment in the *Applications* folder. Follow the steps listed in section 6.1 following »Now open the development environment ...« and test your Arduino board. However, the port is not called *COMxx*, but bears a name like */dev/tty.k610i-SerialPort1*.

On Mac OS X, the Arduino programming environment can be found in the Applications folder

Tip Further information by the Arduino team itself can be found on:

http://arduino.cc/en/Guide/MacOSX

6.3 | Installation on Linux

Tips for the installation on Linux

The installation on Linux is not straightforward and varies depending on the distribution. This is especially true for the driver installation. The following links give you detailed up-to-date information.

Latest notes and important tips can be found on:

http://playground.arduino.cc/Learning/Linux

A very thorough description for Ubuntu can be found on:

http://blog.markloiseau.com/2012/05/installarduino-ubuntu/
http://www.pluggy.me.uk/arduino-ubuntu/

TURN ON YOUR CREATIVITY

FRANZIS
ARDUINO

ARDUINO PROGRAMMING ENVIRONMENT

```
// Blink
// Uses LED L on the Arduino Uno

int ledPin = 13; // LED is connected with digital
pin 13

// The setup routine configures the digital port
// This routine is executed only once
// at program start!

void setup()
{
    // The pin is configured as output
    pinMode(ledPin, OUTPUT);
}
```

7

You have installed the Arduino environment on your computer and successfully carried out a first functional test. The Arduino IDE – short for »integrated development environment«, which just means »programming environment« – provides several tools and settings to simplify the handling of the Arduino.

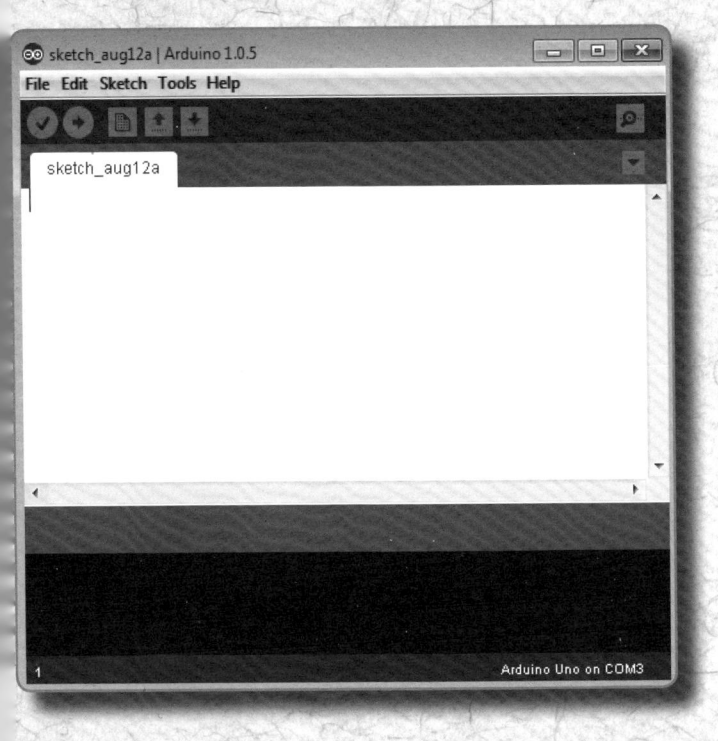

Figure 7.1: The Arduino programming environment: plain and uncluttered

The toolbar of the IDE

Let us have a look at the Arduino programming environment. All features of the IDE can be called via the menus. You will get familiar with the many options while you work with the tutorial kit. Below the menu bar is a tool bar with icons for the most frequently used commands.

Figure 7.2: The menu

Elements of the toolbar
1. Validate and compile program code (syntax check)
2. Upload the program to the Arduino board
3. New (creates a new *sketch editor file*)
4. Open (opens an existing *sketch*)
5. Save (saves the current program)

Info Programs for the Arduino are called *sketches*.

TURN ON YOUR CREATIVITY

FRANZIS
ARDUINO

YOUR FIRST ARDUINO PROGRAM

```c
#include <avr/io.h>
#define F_CPU 1600000UL
#include <util/delay.h>

int main(void)
{
    DDRD  |= (1 << PD0);        // Set PD0 as output
    while(1)                    // Infinite loop
    {
        PORTD |= (1 << PD0);    // PD0 high
        _delay_ms(1000);
        PORTD &= ~(1 << PD0);   // PD0 low
        _delay_ms(1000);
    }
}
```

8

You have already used the program *Blink* to carry out a first functional test. To get familiar with programming, you should begin with writing your own little blink program.

By the way: Don't get discouraged when your first experiments fail! You will learn how to overcome any obstacles step by step und quickly test your own programs.

We recommend the following websites on the topics of electronics and microcontrollers to gain further knowledge online:

http://www.mikrocontroller.net/
http://www.elektronik-kompendium.de/
http://www.roboternetz.de/

Let us now start with the first program. First of all, close any editor windows you might have opened in exploring the development environment except one.

Blink – your first Arduino program

How to write your first program

Storing programs

Time needed: 10 min
Difficulty: 1
■ ☐ ☐ ☐ ☐

Click *File > New*. A new editor window appears. Enter the following program into this window and save it in a folder of your choice on your hard disk. When you specify a name in the Save As dialog, the IDE automatically creates a folder for the program and saves it in that location. For instance, when you specify the program name *LED_blinks*, the programming environment creates a folder with this name in the selected directory and saves the program code in the file *LED_blinks.ino*.

```
// Blink
// Uses LED L on the Arduino Uno

int ledPin = 13; // LED is connected with digital pin 13

// The setup routine configures the digital port
// This routine is executed only once at program start!

void setup()
{
    // The pin is configured as output
    pinMode(ledPin, OUTPUT);
}

// The main program is an infinite loop
void loop()
{
    digitalWrite(ledPin, HIGH); // Turn on LED
    delay(1000);                // Wait for a second
    digitalWrite(ledPin, LOW);  // Turn off LED
    delay(1000);                // Wait for a second
}
```

Example: LED_blinks

Now you can test the program for syntax errors by clicking on the first icon on the toolbar (see Fig. 8.1) or pressing `Ctrl` + `R`.

Caution: Always check the syntax!

Figure 8.1:

When you do not receive any error messages, you can upload the program to the Arduino board by clicking the icon showing an arrow to the right (see Fig. 8.1) or by pressing `Ctrl` + `U`.

Figure 8.2:

When the Upload is completed, the LED L on the Arduino board will begin to blink.

Tip When you receive an error message in the IDE, detaching the board from the PC and then reattaching it may do the trick.

Troubleshooting: If the program does not work

8.1 | What did we do?

The Arduino IDE already did a lot of work for us. Let us have a closer look at our little program:

```
// Blink
// Uses LED L on the Arduino Uno
```

Up to this point, the code only consists of commentaries. The two slashes (//) start a single-line comment.

```
int ledPin=13
```

Here you define the variable `ledPin` with a value of 13. This variable holds the number of the digital pin to which the LED L is attached.

```
void setup()
{
    // The pin is configured as output
    pinMode(ledPin, OUTPUT);
}
```

The pinMode command configures the digital pins as inputs or outputs

The `void setup()` routine is called once at the start of the program and initialized. With the command `pinMode`, you tell the program that pin 13 has to be treated as output. You can configure digital ports as inputs or outputs with this command.

8   Your First Arduino Program   83

```
void loop()
{
    digitalWrite(ledPin, HIGH);  // Turn on LED
    delay(1000);                 // Wait for a second
    digitalWrite(ledPin, LOW);   // Turn off LED
    delay(1000);                 // Wait for a second
}
```

After `void setup()` you see the routine with the actual program. It is called `void loop()` and is called immediately after `void setup()`. This routine is always executed as an infinite loop.

The LED is turned on with the command `digitalWrite`. The parameters in parentheses specify that pin 13 (as identified by the variable `ledPin`) has to be set to `HIGH` (i.e. to be turned on). You could directly type 13 instead of `ledPin`, but that would impair the readability and maintainability of the program.

The next command is `delay`. This command causes a pause. The parameter in parentheses specifies the delay in milliseconds. After this period of time, the LED is turned off by a `digitalWrite` command with the parameter `LOW`. Since we want the LED to be inactive for the same period as it was active, we repeat the command `delay` with 1000 ms. This little program is now continuously executed until you upload another one or cut off the power supply.

Pausing with »delay«

An Arduino program is always structured like this:

```
void setup()
{
    // This code is executed once at the start
    // of the program
}
```

```
void loop()
{
    // The main program runs in this infinite loop
}
```

These two functions must always exist and are essential elements of Arduino programs!

Info The programming language Arduino C uses the hardware abstraction layer (HAL). This makes programming much easier because you do not have to study substantial volumes of data sheets for the microcontroller.

The purpose of the HAL

The HAL is an intermediate logical layer of the operating system which is used by the Arduino programming environment and the underlying Arduino compiler. This intermediate layer attends to the direct hardware initializing. As you can see in the example given above, you only have to enter the command pinMode in order to configure a pin as input or output without needing any detailed knowledge of the microcontroller.

8 Your First Arduino Program

In C, our program may look as follows (this is an example version for Atmel controllers):

```c
#include <avr/io.h>
#define F_CPU 1600000UL
#include <util/delay.h>

int main(void)
{
    DDRD |= (1 << PD0);      // Set PD0 as output
    while(1)                 // Infinite loop
    {
        PORTD |= (1 << PD0); // PD0 high
        _delay_ms(1000);
        PORTD &= ~(1 << PD0); // PD0 low
        _delay_ms(1000);
    }
}
```

TURN ON YOUR CREATIVITY
FRANZIS
ARDUINO

ARDUINO
PROGRAMMING
BASICS

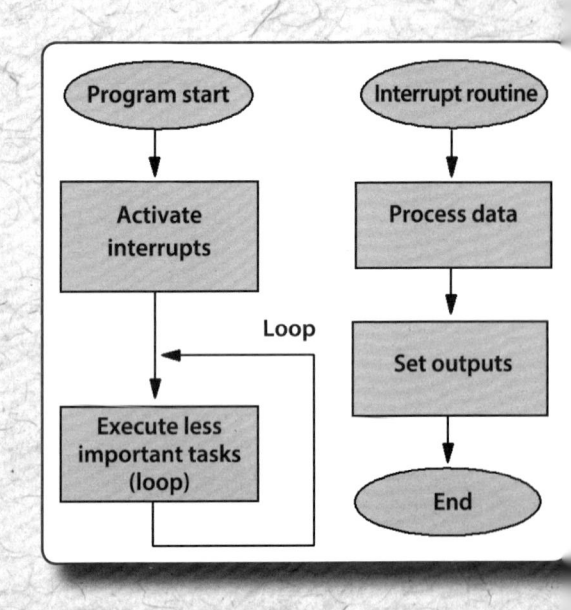

9

Basics of Arduino programming

This chapter provides an introduction to the basics of programming and the programming language *Arduino C*. You will learn how easily and quickly you can write your own programs with just a few commands. Regardless of what types of project you want to carry out with your Arduino, you will find the fundamentals in this chapter.

9.1 | Bits and Bytes

The meaning of »bits« and »bytes«

The smallest unit of information in a computer system is called a *bit*, short for *binary digit*.

A bit can only have one of two different values: 1 or 0 (representing current or no current, respectively). By linking a defined number of bits, you can represent any piece of information.

A set of 8 bits is called a *byte*. A single byte can represent 256 different combinations of 1 and 0. A smaller set of 4 bytes is called a *half byte* or sometimes a *nibble*.

```
4 bits = 1 nibble = 1 half byte
8 bits = 2 nibbles = 1 byte
```

Prefixes for these units correspond to a conversion factor of 1024, not the normal 1000. The reason for this unusual factor is that these are base-2 calculations (2^n).

Conversion formulas for bits and bytes

```
1 byte = 8 bits
1 kilobyte (1 KB) = 1024 bytes (2¹⁰ bytes)
1 megabyte (1 MB) = 1024 KB (2¹⁰ KB)
1 gigabyte (1 GB) = 1024 MB (2¹⁰ MB)
1 terabyte (1 TB) = 1024 GB(2¹⁰ GB)
1 petabyte (1 PB) = 1024 TB (2¹⁰ TB)
```

9.2 | Structure of a Program

Procedural programming

Most programs have similar structures because they follow the pattern of »procedural programming«. Procedural programming refers to the approach of constructing computer programs out of smaller partial problems or tasks that are called procedures, like the function `void setup()` and `void loop()` mentioned above. The smallest step in this approach is a *statement*, like `digitalWrite`.

A program *proceeds* one statement by the other, hence the term »procedural« (derived from the Latin word »procedere«). With his program, the programmer tells the microcontroller what it has to do in which order. This approach is used in order to design the source code in a reusable (modular) and simple way.

9 Arduino Programming Basics

9.2.1 | Sequential Program Flow

In sequential programming, you have always the same code that is executed repeatedly in a loop. It consists of several procedures. The figure shows the basic program flow encompassing input, processing and output.

Basics of sequential programming

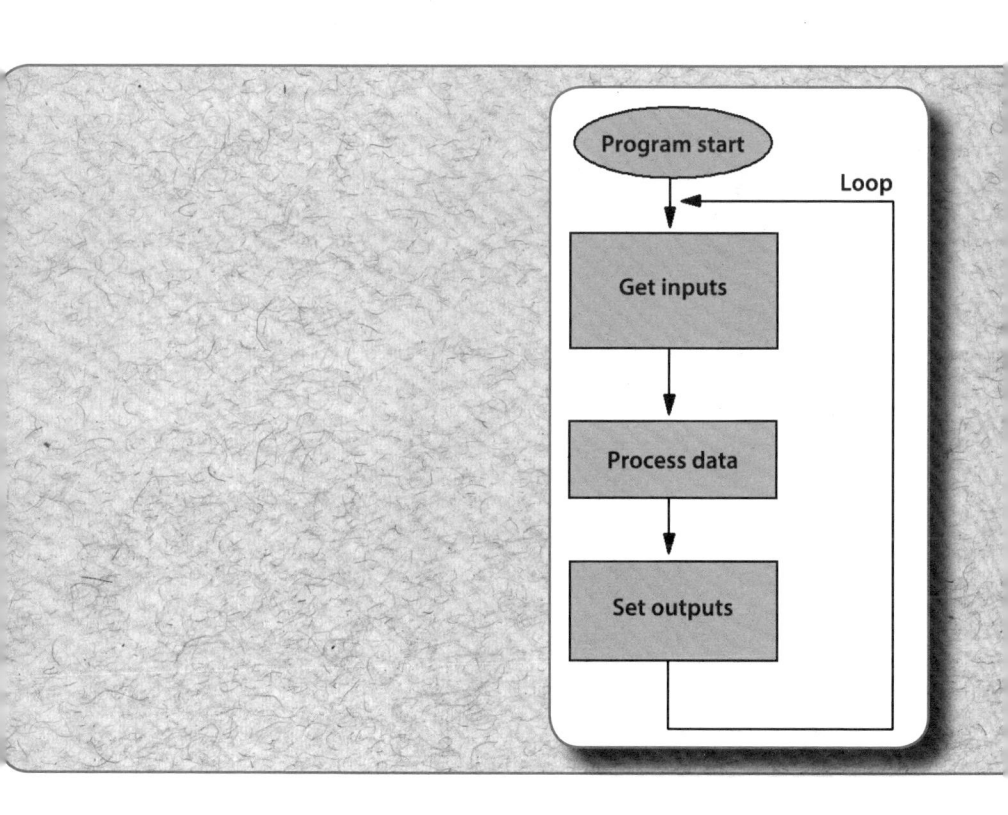

Figure 9.1: Procedural (sequential) program flow

9.2.2 | Interrupt-driven program flow

How does an interrupt-driven program flow work?

In an interrupt-driven program flow, the necessary interrupts (for instance, an input pin that supports interrupts are armed at program start. As with sequential programming, the less important part of the program is executed in an infinite loop like `void loop()`, that runs several procedures. Once an interrupt is fired (for example, by a push-button), the code leaves the infinite loop (called *main loop*) and jumps into the interrupt routine.

Regardless of the position in the program, the current memory state is saved and the interrupt routine is executed. Any important tasks in this routine are now completed, for instance emergency stops or the like. After that, the memory state is reconstructed and the main loop is resumed. You can see this in the following figure.

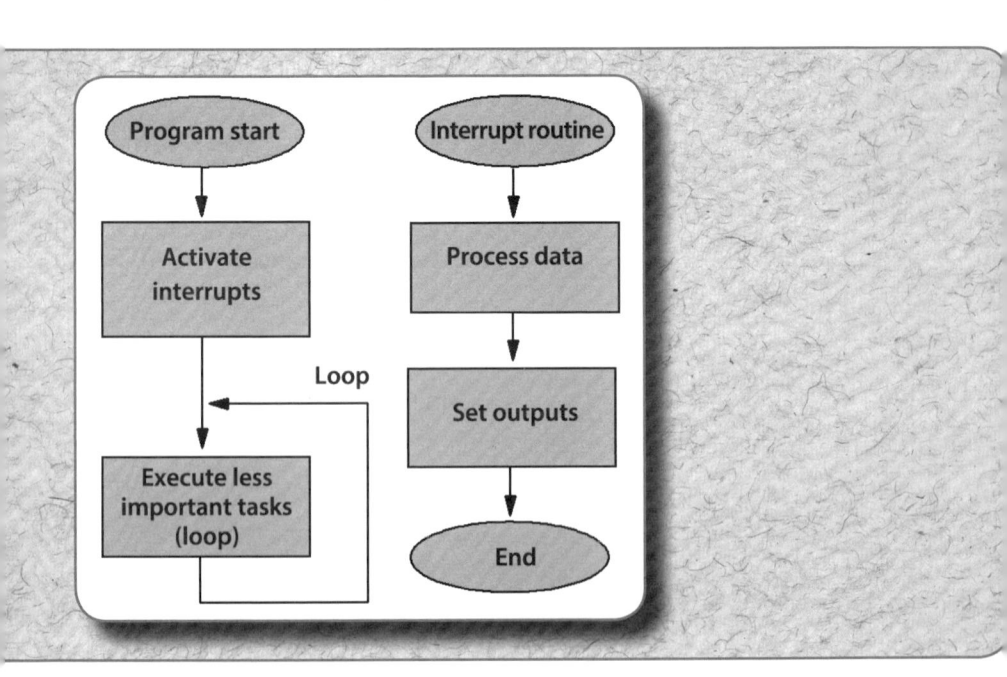

Figure 9.2: Interrupt-driven program flow

9.2.3 | Structure of an Arduino Program

> - Informative text and program description
> - Including libraries
> - Creating global variables
> - Setup routine `void setup()` (initialising pins and variables)
> - Main loop `void loop()`

Structure of an Arduino program

As you can see, it is relatively easy to follow the basic rules.

Informative text and program description

You should always document your programs thoroughly. This also applies to the program header where you give a concise description of the program.

Best practices for Arduino programming

Including libraries

There are ready-made libraries for many complex tasks like Ethernet communication or the handling of servos and sensors. You have to include the appropriate libraries at the beginning of the program before `void setup()`. The functions in these libraries are then accessible in the rest of the program through the respective calls.

Creating global variables

Global variables are variables that are not limited to a certain routine, but can be used at any point in the program. To maintain the clarity of the program, you should not define too many variables with global scope. Furthermore, these variables always take up memory even when they are not used at the moment.

Setup routine void setup() (initialising pins and variables)

Next, you have the familiar routine `void setup()` where you execute several nonrecurring tasks that have to be carried out at program start. `void` means that this function has no return value.

Main loop void loop()

The main routine follows in the form of an infinite loop. Here you can enter the actual code for your program. This function does not return a value either and is therefore of the type `void` as well. We will discuss functions in more detail later.

A second program

Now you have the opportunity to put your newly gained knowledge into practice. With your first program you succeeded in making the LED L on the Arduino board blink. Next, we will extend the program to output a *string* that is sent to the PC via the serial interface of the microcontroller. You can view the output text in a terminal program that is included in the Arduino IDE.

9.3 | Our second Arduino Program

Type the source text manually as an exercise. When you do not want to do that, you can find the code on the CD-ROM in *Samples/Calc*.

After uploading the program to the Arduino, start the Terminal program. You can find it in the Arduino IDE under *Tools*.

9 Arduino Programming Basics

This little program adds two numbers and outputs the result in the terminal. After that, the LED L begins to blink. When you press the reset button, the program starts again.

A little summation program

Time needed: 20 min
Difficulty: 2
■ ■ ☐ ☐ ☐

```
// Franzis Arduino
// My second Arduino program

int ledPin = 13; // The LED is connected to pin 13

void setup()
{
   Serial.begin(9600);
   pinMode(ledPin, OUTPUT);
   Serial.println("Our second Arduino program");
   Serial.println();
}

void loop()
{
   Serial.print("5 + 188 yields ");
   Serial.print(5+188);
   while(true)
   {
      digitalWrite(ledPin, HIGH); // Turn on LED
      delay(1000);                // Wait for a second
      digitalWrite(ledPin, LOW);  // Turn off LED
      delay(1000);                // Wait for a second
      continue;
   }
}
```

The program already shows the main features of a sequential program. In the first step, the Arduino is told that the `ledPin` variable contains the number 13 (as in the first program). Next, the program enters the setup routine. Here you define the pin to which the LED is connected and a transfer rate of 9600 baud for the serial interface. At the end of the setup routine, information is output via the serial interface.

Next we have the main loop `void loop()` where we will output a text and the result of 5 + 188. In order to display these pieces of information in the same line, you use the command `Serial.print`. We used `Serial.println` for the first text that was output via the serial interface. This causes a combined carriage return and line feed (*CR+LF*) at the end of the line.

After the output, the program enters the `while` loop, where it is trapped until you stop it by uploading a new program or resetting the board. By continuous blinking, the LED signals that the program has reached the end.

Implications of an infinite loop

As you can see, an infinite loop causes a state that you cannot cancel without external influence.

Tip When entering the code, pay attention to parantheses, semicolons and the correct spelling (upper/lower case) of programming commands.

9.4 | Getting Started with Arduino Programming

As you are a beginner in Arduino programming, this chapter is intended to make you familiar with the most important programming fundamentals, the basic commands and functions of the Arduino programming language. For the following programming steps, it is important that you study this chapter thoroughly. Try out the examples and make sure you understand them. Feel free to experiment with minor modifications. As always, practice makes perfect. What you now practice, you will not have to look up later.

Learning a new programming language is like learning the vocabulary of a foreign language. The only difference is that the programming environment will mercilessly point out any mistakes you make. When the code is wrong, the program will not function correctly.

Basic commands and functions

9.4.1 | Commentaries

If you want to be able to read and understand your programs after some time, you have to document the source text clearly and explicitly. According to experience, even the most seasoned programmers can no longer remember every detail of a previous project after some weeks have elapsed or many other projects had to be attended to. Thus, it is very important to document the program code with additional important information. This documentation does not have to be saved in a separate file, but can be noted in the source code itself at the appropri-

Why it is important to document your programming activities

ate positions. To this end, there are several symbols that mark text as commentary, either as single-line or block commentary. Commentaries are always depicted with a gray background in the Arduino IDE.

Example: Commentaries

Time needed: 5 min
Difficulty: 1
■ ☐ ☐ ☐ ☐

```
// This is a single-line commentary.

/*
This is a
multi-line commentary
that gets longer and longer and longer ...
*/

/*****************************************
Now this looks much better!
*****************************************/
```

9.4.2 | Braces

The meaning of braces

Braces ({ }) tell the compiler that there is a block of code. A block always begins with { and ends with }. Between these braces, the statements are given. Think of the compiler as some sort of translator that converts program code to machine code, so that the microcontroller can understand and process it.

Tip More information about compilers can be found at:

http://en.wikipedia.org/wiki/Compiler

Example:

```
void My_function()
{
    // Commands are placed inside the braces
}
```

9.4.3 | Semicolon

The semicolon (;) completes a statement. If you omit this symbol, the compiler will immediately issue one or more error messages. Try to run one of the previous examples with one semicolon missing!

Do not forget the semicolon

Example:

```
int x=42;;
```

Here, x is defined as an *integer variable* and assigned the value 42. The semicolon completes the assignment.

9.4.4 | Data types and variables

Typically, a program encompasses several variables that are used either to get values from the external world (e.g. from an analog input or a digital pin) or to facilitate internal calculations. For programming purposes, there are several variable types like `byte`, `inter`, `long`, and `float`. All variables must be defined before use.

Variable types »byte«, »integer«, »long«, and »float«

Example:

```
byte myVariable = 0;
```

A variable of the type `byte` can hold numerical values from 0 to 255 (1 byte). Here, the variable `myVariable` is created and initialized with 0. Afterwards you can assign values to `myVariable` by using =. When you want to pass the result of 3 x 3 to this variable, you write the following:

```
myVariable = 3*3;
```

9.4.5 | Variable names

Variable names are case sensitive

Variable names are case sensitive. When you create a variable named `myVariable`, you have to refer to it by using exactly this spelling without any variations. The compiler will not understand the alternative version `MyVariable`.

Composite variable names

An underscore (_) is allowed as a separator symbol for variable names that are made up of several elements. You could also define a variable as follows:

```
byte my_Variable = 0;
```

The underscore is often used to make long variable names much more readable.

Important Key words like `if`, `while`, `do` etc. are not allowed as variable names. Global variables and functions must not bear the same names. Furthermore, when a function has the same name as a local variable, it can not be used in the scope of that variable.

9.4.6 | Local and global variables

When you declare a variable inside a function or as the argument of a function, it is scoped locally. This means that the variable only exists in the scope of this function. Variables declared outside of a function are called global variables. We have seen variables like this before. They are accessible to all functions of the program.

Variables are either locally or globally scoped

```
byte Variable;          // A variable of the type byte that can hold values
                        // from 0 to 255
float PI = 3.1415;      // The constant PI as a float variable
int myArray[10];        // A byte array. This resembles creating ten times a
                        // byte with "byte Var;". You refer to the specific
                        // bytes with the index: "Var (x)".
                        // Numbering of arrays
                        // begins with 0. In this case, the ten variables are
                        // numbered from 0 to 9.
```

9.4.7 | Usage of the different data types

The following descriptions show the different data types and specify the amount of memory they use.

boolean

Variables of the type `boolean` can have one of the two states `true` and `false`. A `boolean` variable uses up 1 byte of memory.

List of data types

```
boolean MyTruth = true;   // The variable is true
boolean MyTruth = false;  // The variable is false
```

byte
1 byte equals 8 bits. A byte variable can hold values from 0 to 255.

```
byte MyVariable = 0;
    // The variable is created and initialized with 0
```

char

Characters and unsigned characters

A *character* (a symbol) comprises 1 byte. A char variable contains a character set in *single quotes* that is saved as the corresponding number in the ASCII character set, e.g. 65 for the letter A. char variables can hold values between -127 and +127.

```
char MyCharacter = 'A'; // Saved as 65
```

unsigned char
Unsigned characters resemble the normal signed characters, but can hold only positive values from 0 to 255.

```
unsigned char MyCharacter = 'B'; // Saved as 66
```

int
An *integer* comprises 2 bytes and can hold values from -32,768 to +32,768.

```
int MyVariable = -32.760;
    // Integer variable holding the value -32.760
```

unsigned int

Unsigned integer variables can hold values from 0 to 55,535 (2^{16} - 1). In opposition to `int`, this data type is not signed (does not allow for negative values). It uses up the same amount of memory, i.e. 2 bytes.

```
unsigned int MyVariable = 50.000;
    // Integer variable holding the value 50.000
```

long

A `long` variable uses up 4 bytes and can hold values from -2,147,483,648 to +2,147,483,647 (32-bit `long`).

long and unsigned long

```
long MyVariable = 10.000.000;
    // Integer variable holding the value
    // 10.000.000
```

unsigned long

An `unsigned long` variable uses up 4 bytes and can hold values from 0 to 4,294,967,295 (2^{32} -1; 32-bit `unsigned long`). It is not signed and therefore does not allow for negative values.

```
unsigned long MyVariable = 54.544.454.544;
    // Very big variable
```

float

Float variables can hold signed 32-bit values from a range between -3.4028235 × 10^{38} and +3.4028235 × 10^{38}. They use up 4 bytes of memory.

```
float MyVariable = 100.42;
    // A float variable holding the value 100.42
```

Strings

A *string variable* is a combination (an array) of `char` variables plus a terminating symbol.

```
char MyString[] = "Hello World";
    // You need 12 bytes
```

Arduino operators and their effect

Every single character needs 1 byte, and at the end of the string, another byte is used for terminating. If your strings only consist of the word »hello«, you need 6 bytes.

Arrays

An *array* is a collocation of variables in the memory. It is regarded as a data structure in computer sciences. With arrays, you can place data – typically of uniform data type (`byte`, `int` etc.) – in an organized way inside the memory so you can access it via an index. In the Arduino IDE, arrays have to be declared as having the data type `int`.

Numbering of array indices begins with 0

In the Arduino IDE, the numbering of array indices begins with 0. (Some other compilers, for instance many Basic compilers, use 1 as the first number.)

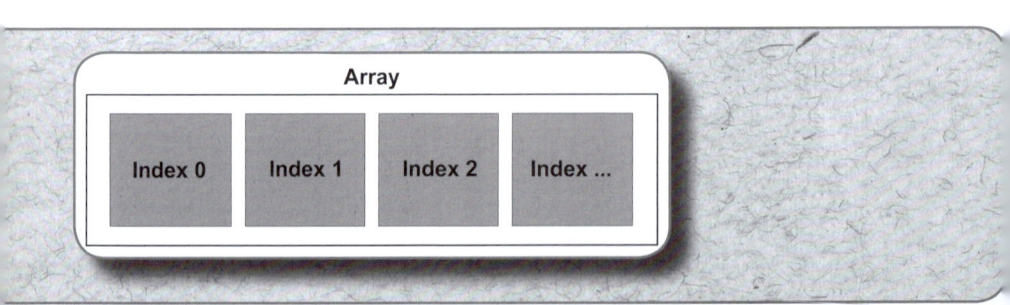

Figure 9.3: Organization of an array in memory. Each index represents a variable and thus a saved value that can be written and read by referring to the index.

Example: Arrays

Time needed: 10 min
Difficulty: 1

■ ☐ ☐ ☐ ☐

```
// Franzis Arduino
// Arrays

int Array_1[3];
int Array_2[] = {1,2,3};

void setup()
{
   Serial.begin(9600);
   Serial.println("Arduino Arrays");
   Serial.println();
}

void loop()
{
   byte x;

   Array_1[0] = 1;
   Array_1[1] = 2;
   Array_1[2] = 3;

   Serial.println("Output of array 1 ");
   Serial.println("-----------------");

   // Output the data of the first array
   for(x=0;x<3;x++)
   {
      Serial.print(Array_1[x]);
      Serial.println();
   }

   Serial.println("Output of array 2 ");
   Serial.println("-----------------");
```

```
// Output the data of the second array
Serial.print(Array_2[0]);
Serial.println();
Serial.print(Array_2[1]);
Serial.println();
Serial.print(Array_2[2]);

while(1);
}
```

This example shows the use of arrays. We preset a size of 3 for the first array, thus we can save three `int` Variables (16 bit) in it. The second array is a *dynamical array*: It has no predetermined size, but we deposit some values inside the braces. In the main loop we pass values to the first array, in this case 1, 2 and 3. You can use your own values here in order to gain a better understanding of arrays. The counter variable x in the `for` loop iterates through the index of the array and outputs the values in Terminal. The values of the second array are selectively retrieved by referring to the index and output in Terminal as well.

The program remains in the infinite `while(1)` loop. That way, it is guaranteed that it is only executed once.

9.4.8 | Operators

Every data type has its specific operators that define which operations can be executed on it. The following list shows all the available Arduino operators and their use.

Arithmetical operators
- =   Result or carry-over
- +   Summation
- -   Subtraction

*   Multiplication
/   Division
%   Modulo, i.e. remainder

Comparison operators
== Equals, e.g. A == B
!= Not equal, e.g. A != B
< Less than, e.g. A < B
> Greater than, e.g. A > B
<= Less than or equal to, e.g. A <= B
>= Greater than or equal to, e.g. A >= B

Comparison bitwise arithmetic and Boolean arithmetic

Bitwise arithmetical operators
& Bitwise AND
| Bitwise OR
~ Bitwise NOT; reverses a bit

Boolean arithmetical operators
&& AND, e.g. if Answer_A && Answer_B are true, then do something
|| OR, e.g. if Answer_A || Answer_B are true, then do something
! NOT, e.g. if ! Answer_A is true, then do something

Increment and decrement operators
++ Increment, e.g. I++ to increase variable I by 1
-- Decrement, e.g. I-- to decrease variable I by 1
+= Increment, e.g. I +=5 to increase variable I by 5
-= Decrement, e.g. I -=5 to decrease variable I by 5
*= Multiplication, e.g. I *=2 to multiply variable I by 2
/= Division, e.g. I /=2 to divide variable I by 2

Constants
HIGH/LOW   HIGH = 1, LOW = 0
INPUT/OUTPUT   INPUT = 0, OUTPUT = 1
true/false   true = 1, false = 0

Constants

9.4.9 | #define statements

 # (hash sign)

#define is a preprocessor statement that is executed before compilation. You can think of it as some sort of parse that converts the #define statements to constants. This makes it possible to assign a constant value to a name. However, these are not classical constants, but values passed to a #define expression.

Tip More information about preprocessors can be found at:

http://en.wikipedia.org/wiki/Preprocessor

Example:

```
#define myDefine 1
    //without semicolon and "=" sign!
```

Here, the value 1 is assigned to the expression myDefine. Every occurrence of the name myDefine in the program is then replaced by 1. This happens invisibly.

9.4.10 | Control structures

In order to respond to events, every program needs a way for the conditional execution of code. In C, these so-called control structures are if, else if, else or switch case.

The output of the following sample programs is again shown in the Terminal program that comes with the Arduino IDE. Upload the examples to the board and start Terminal in order to see the results.

if

```
if(VariableA == VariableB)
{
    // This code is executed when VariableA
    // equals VariableB
}
```

Example: if

```
// Franzis Arduino
// if…

int x;

void setup()
{
    Serial.begin(9600);
    Serial.println("if statements");
    Serial.println();
}

void loop()
{
    if(x==10)
    {
        Serial.println("The counter of variable X
            now has a value of 10!");
        while(1);
    }
    x++;
}
```

Time needed: 10 min
Difficulty: 1

if implements branching of programs

The program code runs in the main loop `void loop()` until the integer variable `x` becomes `10`. Only now the part of the program between the braces of the `if` statement is executed. This way, you can easily implement a branching in your programs. Try to experiment with other operators! Any logical comparisons like `!=`, `<` and `>` are allowed. Next, we will look at the variant `if ... else`.

if ... else

```
if(VariableA > VariableB)
{
    // This code is executed when the condition
    // is true
}
else
{
    // Applies when VariableA is not greater
    // than VariableB
{
    // This code is executed when the condition
    // is not true
}
```

Example: if ... else

Time needed: 10 min
Difficulty: 1
■ ☐ ☐ ☐ ☐

```
// Franzis Arduino
// if ... else

int x;

void setup()
{
    Serial.begin(9600);
```

```
    Serial.println("if ... else statements");
    Serial.println();
}

void loop()
{
    x++;
    if(x>10)
    {
        Serial.print("The counter of variable X
            has a value of ");
        Serial.println(x);
        while(1);
    }
    else
    {
        Serial.print("X = ");
        Serial.print(x);
        Serial.println();
    }
}
```

With else, you can offer an alternative. The program increments the value of x and shows it as long as it is less than or equal to 10. Only when x becomes greater than 10, is the code in the braces of the if statement executed. The operands are the same as in the previous example.

else if

else if offers another possibility for nesting several if statements. With this keyword, you can ask for several diverse states of the variable. Depending on whether the condition is true or false, the corresponding block in the else if statement is executed.

else if

```
if(VariableA == VariableB)
{
    // This code is executed when VariableA
    // equals VariableB
}
else if (VariableA > VariableB)
{
    // This code is executed when VariableA is
    // greater than VariableB
}
else
{
    // This code is executed when VariableA is
    // less than VariableB
}
```

In the following example, different text is output depending on the value of variable x.

Example: else if

Time needed: 10 min
Difficulty: 2

```
// Franzis Arduino
// else If

int x;

void setup()
{
    Serial.begin(9600);
    Serial.println("else if statements");
    Serial.println();
}

void loop()
{
```

```
if(x==1)
{
    Serial.println("Text 1");
}
else if(x==10)
{
    Serial.println("Text 10");
    while(1);
}
else
{
    Serial.print("X = ");
    Serial.print(x);
    Serial.println();
}
x++;
}
```

switch case

The `switch case` statement resembles `else if`. They both have a specific code block that is executed depending on which condition is `true`. Optionally, you can provide a standard alternative with `default` as you did it with `else`. If none of the `case` conditions hold true, `default` is executed. Every `case` statement ends with `break`. At this point the program leaves the `case` block.

The switch case statement and its use

```
switch(Variable)
{
    case 1:
    // This code is executed when Variable = 1
    break;
```

```
    case 2:
    // This code is executed when Variable = 2.
    break;

    default:
    // This code is executed when both conditions
    // are false
}
```

Example: switch case

Time needed: 10 min
Difficulty: 2
■ ■ ☐ ☐ ☐

```
// Franzis Arduino
// Switch Case

int x;

void setup()
{
    Serial.begin(9600);
    Serial.println("switch case statements");
    Serial.println();
}

void loop()
{
    switch(x)
    {
        case 10:
        Serial.println("We have reached 10");
        break;

        case 20:
        Serial.println("We have reached 20");
        break;
```

```
case 30:
  Serial.println("We have reached 30");
  while(1);
  break;

default:
  Serial.print("X = ");
  Serial.print(x);
  Serial.println();
}

x++;
}
```

The program runs through the main loop and increments variable x by 1. Depending on which of the cases is true, a corresponding line is printed to Terminal.

9.4.11 | Loops

In programming you often need loops to count (in decimal or in binary numbers), to implement the main loop, or to read from the serial interface as long as there are characters in the buffer, just to name a few of the possibly uses. There are several different types of loop and each has its own characteristics.

Types of loops you should know

for

In a `for` loop, a variable is increased or decreased by a specified increment inside a specified range.

```
for(start condition; stop condition;
    incrementation statement)
{
    // Program block
}
```

The loop in the following example counts from 0 to 10 in increments of 1. As x++ has the same meaning as x=x+1, the variable is incremented. For an increment of 2 you have to write x=x+2.

```
for(x = 0 ; x < 10 ; x++)
{
    // Code that has to run 10 times
    // The variable x is incremented from 0 to 9
    // and acts as a counter
}

// Now the variable x is incremented by 2
for( x = 0 ; x < 10 ; x=x+2)
{
    // Code that has to run 5 times
    // The variable x is incremented from 0 to 8
}
```

9 Arduino Programming Basics

In the next example, the variable x is decremented from 10 to 1 (with a decrement of 1). Here we use x-- which is equal to x=x-1.

```
for(x = 10 ; x > 0 ; x--)
{
    // Code that has to run 10 times
    // The variable x is decremented from 10 to 1
}
```

Example: for

```
// Franzis Arduino
// for

int x;

void setup()
{
    Serial.begin(9600);
    Serial.println("for statements");
    Serial.println();
}

void loop()
{
    Serial.println("Increment 1");
    for(x=0;x<10;x++)
    {
        Serial.print("X = ");
        Serial.print(x);
        Serial.println();
    }
```

Time needed: 10 min
Difficulty: 2

■ ■ ☐ ☐ ☐

```
  Serial.println("Increment 2");
  for(x=0;x<10;x=x+2)
  {
    Serial.print("X = ");
    Serial.print(x);
    Serial.println();
  }

  Serial.println("Now we count down from 10 to 1
    with a decrement of 1");
  for(x=10;x>0;x--)
  {
    Serial.print("X = ");
    Serial.print(x);
    Serial.println();
  }

  // End of program!
  while(1);
}
```

The program passes through the three `for` loops and demonstrates its functionality by printing the counter values.

while and do while

Two further variants: »while« and »do while«

Other loop types are `while` and `do while`. With the `do while` variant, the code in the loop is executed first, and then the condition is tested. Thus this type of loop is executed at least once. Infinite loops are often implemented by `while`. They can be aborted by `break`. When you want to test the condition before passing through the loop, you have to use `while(statement)`. In this variant, the condition is tested before executing the loop.

```
while(1)
{
    // Infinite loop
    // Code for whatever you want to do
}
```

```
while(1)
{
    // Infinite loop with conditional abort
    Variable++;
    if( Variable > 10 ) break;
}
```

```
while(Variable < 10)
{
    // Tests the condition at the beginning
    Variable++;
}
do
{
    // Tests the condition after running through
    // the loop
    Variable++;
}while( Variable < 10 );
```

Loops always test if the condition is `true` (which equals 1). As long as the condition is `true`, the program runs inside the loop.

Example: do while

Time needed: 10 min
Difficulty: 2

```
// Franzis Arduino
// do while

int i=0;

void setup()
{
  Serial.begin(9600);
  Serial.println("while and do while program");
  Serial.println();
}

void loop()
{
  while(1)
  {
    i++;
    Serial.print(i);
    Serial.println();
    if(i>9) break;
  }

  i=0;
  Serial.println();

  while(i<10)
  {
    i++;
    Serial.print(i);
    Serial.println();
  }

  i=0;
  Serial.println();
```

```
    // do while
    do
    {
        i++;
        Serial.print(i);
        Serial.println();
    }while(i < 10);

    while(1);
}
```

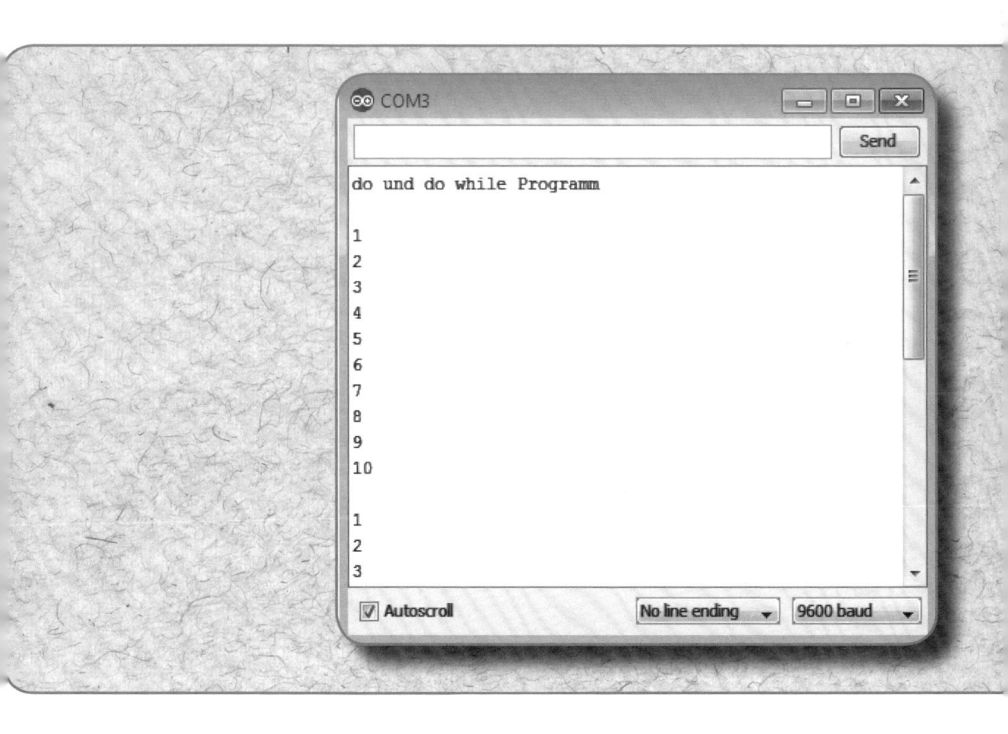

Figure 9.4: The output in Terminal. The Baud rate of the program is set to 9600. The same value has to be selected in Terminal.

Alert: programming with loops

Caution Be alert when programming with loops. Loop variables can easily have an overflow outside of their range. Furthermore, an infinite loop or an unsatisfiable condition can cause the program to hang. The program gets stuck at this point until it is interrupted by an external reset. Therefore, you have to double-check the parts of your programs that contain loops. Print the counter values to Terminal in order to monitor the variables.

9.4.12 | Functions and routines

Functions provide more clarity

Functions significantly improve the clarity of a program and allow you to implement your own commands. You can write your own functions for tasks that you have to execute repeatedly and re-use them in other projects (modularity). In a function, you can for instance carry out a mathematical calculation by passing two numbers to it. The function then responds by returning the result, e.g. the product of the two numbers. In this case, the return type is no longer void, but for example int if you want to return an integer variable.

The following example demonstrates the simplest form of a function with no parameters and no return value. In programming languages like Basic, this type of function is called a subroutine.

Example: Functions I

Time needed: 10 min
Difficulty: 1
■ □ □ □ □

```
// Franzis Arduino
// Functions

void setup()
{
    Serial.begin(9600);
    Serial.println("Arduino functions");
```

```
    Serial.println();
}

void loop()
{
    Output1();
    Output2();

    while(1);
}

void Output1()
{
    Serial.println("Output 1");
}

void Output2()
{
    Serial.println("Output 2");
}
```

In `loop()`, `Output1()` is called at first. The program jumps to this function and sends the output via the serial interface to Terminal on the PC. After that, the program leaves `Output1()` and returns to `loop()`, more precisely to the point after the call to `Output1()`. Now the action is repeated for `Output2()`. Finally, the program reaches `while(1)`, where it stays.

A function is a block of program code that executes several statements and then returns to the initial program. During this process, the function can pass values to other functions or variables. It is useful to write your own functions in order to simplify repetitive tasks and to clarify the program structure. To create a function, you first define its type. In our previous example, the type was `void`, because the function did not return a value.

A function is a block of program code that executes statements

A function with parameters and return value has the form *return_type name* (*parameters*):

```
int myFunction(byte var1, byte var2)
{
    // program code
    return var1 + var2;
}
```

The function is called as follows:

```
Serial.println(myFunction(1,1));
```

The result of calculating the sum of 1 + 1 is directly printed to Terminal. Instead of using `Serial.println` you can also pass the value to a variable:

```
var_add = myFunction(1,1);
```

Now the result is saved in the variable `var_add` (which has of course to be big enough to hold that value!).

Examples for adding two numbers und printing the result to Terminal

The following example shows a simple function that adds two numbers and prints the result to Terminal. You can use any numbers as long as the sum does not exceed the range of integer variables. The `return` statement at the end of the function provides the result. It is also possible to compare two numbers inside a function and then return `true` or `false`. Feel free to experiment with functions. For example, you could program a function that solves a complex arithmetic problem and prints the result to Terminal.

Example: Functions II

```
// Franzis Arduino
// Arduino functions

void setup()
{
  Serial.begin(9600);
  Serial.println("Arduino functions");
  Serial.println();
}

void loop()
{
  int val;
  int x = 12;
  int y = 55;

  val=Add(x,y);
  Serial.print("The sum of ");
  Serial.print(x);
  Serial.print(" + ");
  Serial.print(y);
  Serial.print(" = ");
  Serial.print(val);

  while(1);
}

int Add(int val_1, int val_2)
{
  return val_1 + val_2;
}
```

Time needed: 10 min
Difficulty: 3

■ ■ ■ ☐ ☐

9.4.13 | continue

The continue statement skips the rest of the loop code

The `continue` statement skips the rest of the code in a `do`, `for` or `while` loop and executes the code after the `{ }` block. In the following example, `continue` aborts the `for` loop, when the variable `i` is not divisible by 2.

Example: continue

Time needed: 10 min
Difficulty: 1
■ ☐ ☐ ☐ ☐

```
// Franzis Arduino
// continue

int i=0;

void setup()
{
  Serial.begin(9600);
  Serial.println("The continue statement");
  Serial.println();
}

void loop()
{
  for(i=0;i<10;i++)
  {
    if(i%2==0)
    {
      continue;
    }
    Serial.print(i);
    Serial.print(" is not divisible by 2!");
    Serial.println();
  }

  while(1);
}
```

9.4.14 | Type conversions

With the functions `char()`, `byte()`, `int()`, `long()`, and `float()` you can convert any variable to the specified type. For instance, you can turn a `byte` variable into a `long` variable. This is done to adjust the data type for subsequent calculations of functions.

char()
Converts a value to a character.

byte()
Converts a value to a byte.

int()
Converts a value to an integer.

long()
Converts a value to a long number.

float()
Converts a value to a float number.

The functions char(), byte(), int(), long(), and float()

There may be several reasons for changing the data type of a value. For example, you may have to do subsequent calculations with a higher precision, or you do not need the decimal places any more. In cases like these you carry out a type conversion.

```
int i 0 0;
long(i);
```

Here the integer variable `i` is converted to the data type `long`.

9.4.15 | Mathematical functions

Mathematical functions of the Arduino IDE

In the following section you will be introduced to the mathematical functions of the Arduino IDE. Feel free to verify the results with a calculator.

min(x, y)

min(x, y) and max(x, y)

`min(x, y)` calculates the minimum of two values of the same data type and returns the smaller one.

Example: min

Time needed: 5 min
Difficulty: 1
■ ☐ ☐ ☐ ☐

```
// Franzis Arduino
// min(x,y) function

int x,y,Res=0;

void setup()
{
   Serial.begin(9600);
   Serial.println("min(x,y) function");
   Serial.println();
}

void loop()
{
   Res=min(10,55);
   Serial.print(Res);
   Serial.println();

   while(1);
}
```

max(x, y)

max(x, y) calculates the maximum of two values of the same data type and returns the greater one.

Example: max

```
// Franzis Arduino
// max(x,y) function

int x,y,Res=0;

void setup()
{
    Serial.begin(9600);
    Serial.println("max(x,y) function");
    Serial.println();
}

void loop()
{
    Res=max(10,55);
    Serial.print(Res);
    Serial.println();

    while(1);
}
```

Time needed: 5 min
Difficulty: 1
■ ☐ ☐ ☐ ☐

Further mathematical functions

Time needed: 5 min
Difficulty: 1

abs(x)

abs(x) calculates the absolute value of the argument.

Example: max

```
// Franzis Arduino
// abs(x) function

int Res;

void setup()
{
   Serial.begin(9600);
   Serial.println("abs(x) function");
   Serial.println();
}

void loop()
{
   Res=abs(3.1415);
   Serial.print(Res);
   Serial.println();

   while(1);
}
```

constrain(x, a, b)

`constrain(x, a, b)` constrains the number `x` to the range from `a` to `b`.

Example: max

```
// Franzis Arduino
// constrain(x, a, b) function

int x,Res;

void setup()
{
    Serial.begin(9600);
    Serial.println("constrain(x, a, b) function");
    Serial.println();
}

void loop()
{
    for(x=0;x<60;x++)
    {
        Res=constraint(x, 10, 50);
        Serial.print(Res);
        Serial.println();
    }

    while(1);
}
```

Time needed: 5 min
Difficulty: 1

■ □ □ □ □

map(x, fromLow, fromHigh, toLow, toHigh)

The map function is used to scale numbers from one range to another

The map function is very useful to map one range to another. This is ideally suited to scale a great input variable to a smaller output variable.

Example: max

Time needed: 5 min
Difficulty: 1

```
// Franzis Arduino
// map(x, fromLow, fromHigh, toLow, toHigh) function

int x,Res;

void setup()
{
    Serial.begin(9600);
    Serial.println("map(x, fromLow, fromHigh,
        toLow, toHigh) function");
    Serial.println();
}

void loop()
{
    for(x=0;x<20;x++)
    {
        Res=map(x,0,20,5,15);
        Serial.print(Res);
        Serial.println();
    }

    while(1);
}
```

pow(base, exponent)

The pow function returns the result of raising the first argument to the power of the second one. In the following example, both arguments and the result are of the type float.

An example for the pow function

Example: pow

```
// Franzis Arduino
// pow(base, exponent) function

float Res;

void setup()
{
    Serial.begin(9600);
    Serial.println("pow(base, exponent) function");
    Serial.println();
}

void loop()
{
    Res=pow(10,5);
    Serial.print(Res);
    Serial.println();

    while(1);
}
```

Time needed: 5 min
Difficulty: 1

sq(x)

The argument is multiplied with itself (x * x).

The function for squaring a number

Example: sq(x)

Time needed: 5 min
Difficulty: 1
■ □ □ □ □

```
// Franzis Arduino
// sq(x) function

int Res;

void setup()
{
   Serial.begin(9600);
   Serial.println("sq(x) function");
   Serial.println();
}

void loop()
{
   Res=sq(3);
   Serial.print(Res);
   Serial.println();

   while(1);
}
```

sqrt(x)

`sqrt(x)` calculates the square root of a number. This is the counterpart of `sq()`.

How to calculate a square root

Example: sqrt(x)

Time needed: 5 min
Difficulty: 1
■ □ □ □ □

```
// Franzis Arduino
// sqrt(x) function

int Res;
```

9 Arduino Programming Basics

```
void setup()
{
    Serial.begin(9600);
    Serial.println("sqrt(x) function");
    Serial.println();
}

void loop()
{
    Res=sqrt(9);
    Serial.print(Res);
    Serial.println();

    while(1);
}
```

sin(rad)

Calculates the sine of the angle specified in radian. The return value is the sine of the input value (in the range from -1 to +1).

Trigonometric functions

Example: sin(x)

```
// Franzis Arduino
// sin(x) function

float Res;

void setup()
{
    Serial.begin(9600);
    Serial.println("sin(x) function");
    Serial.println();
}
```

Time needed: 5 min
Difficulty: 1
■ □ □ □ □

```
void loop()
{
   Res=sin(1.0);
   Serial.print(Res);
   Serial.println();

   while(1);
}
```

cos(rad)
Calculates the cosine of the angle specified in radian. The return value is the cosine of the input value (in the range from -1 to +1).

Example: cos(x)

Time needed: 5 min
Difficulty: 1
■ □ □ □ □

```
// Franzis Arduino
// cos(x) function

float Res;

void setup()
{
   Serial.begin(9600);
   Serial.println("cos(x) function");
   Serial.println();
}

void loop()
{
   Res=cos(1.0);
   Serial.print(Res);
   Serial.println();

   while(1);
}
```

an(rad)

Calculates the tangent of the angle specified in radian.

Example: tan(x)

Time needed: 5 min
Difficulty: 1

```
// Franzis Arduino
// tan(x) function

float Res;

void setup()
{
    Serial.begin(9600);
    Serial.println("tan(x) function");
    Serial.println();
}

void loop()
{
    Res=tan(1.0);
    Serial.print(Res);
    Serial.println();

    while(1);
}
```

9.4.16 | Serial communication

In order to transfer data to a computer or another microcontroller, the microcontroller on the Arduino board comes with a UART (Universal Asynchronous Receiver Transmitter). This is an electronic circuit for implementing digital serial interfaces. The communication UART is very useful and versatile. The Arduino IDE provides several com-

The UART and its use

mands for it. Some of them have already been described in previous examples, like `Serial.print()` and `Serial.println()`.

The microcontroller has an integrated UART hardware interface. This means that data can be sent via a dedicated interface inside the microcontroller. The processor does not need to take care of this, and the data transfer does not put any strain on it. It is also possible to emulate the UART in the software. A software UART is not as fast a a hardware UART and needs a decent amount of processing power, but it allows for simultaneous connections to several receivers. Generally, Arduino programs use the hardware UART.

The Arduino command `Serial.println` that we have used in several previous examples sends a string via the hardware UART (the serial interface). The nonprintable characters *CR* (carriage return) and *LF* (line feed) that cause a line break are automatically appended to this string. If you do not want to have a line break (for instance, if you just want to string together several values, you use `Serial.print`.

A number passed to `Serial.print` is automatically converted to a text string by the Arduino IDE. A number like 128 consists basically of three single characters, namely 1, 2 and 8. Thus, not the number itself is sent but the ASCII codes for the individual characters.

Sending ASCII codes to Terminal

It is also possible to transfer ASCII code to Terminal by sending a single byte with the appropriate code. The number 65 correlates to the ASCII code of the letter A (see the ASCII table in the appendix). When you use `Serial.write(65)` to transfer a byte with the value of 65 via the serial interface, the character A is displayed in Terminal (because Terminal uses ASCII coding).

9 Arduino Programming Basics | 137

Serial.begin(*baud rate***)**
In order to use the serial interface, you have to preset its transfer rate with the function `Serial.begin()`. The unit of measurement for this speed is called *baud*. 1 baud means a transfer of one character per second.

Tip More information on the transfer rate can be found at:

http://en.wikipedia.org/wiki/Baud

`Serial.begin(`*baud rate*`)` opens the serial interface (UART) and sets its data transfer rate.

`Serial.begin(`*baud rate*`)`

The following transfer rates are possible:

- 300
- 4200
- 2400
- 4800
- 9600
- 14,400
- 19,200
- 28,800
- 38,400
- 57,600
- 115,200

Possible transfer rates

When you use an Arduino Mega board or an equivalent model with several UART hardware interfaces, the configuration may look as follows:

```
Serial.begin(9600);
Serial1.begin(38400);
Serial2.begin(19200);
Serial3.begin(4800);
```

The individual interfaces are referred to by a sequential number after `Serial`. The only exception is the first interface UART0, where no additional number is specified in configuration. In order to print characters, you have to use `Serialx.print()`, where x represents the number of the interface.

```
Serial.println("Hi, this is UART0!");
Serial1.println("Hi, this is UART1!");
Serial2.println("Hi, this is UART2!");
Serial3.println("Hi, this is UART3!");
```

Info It is not possible to use the digital pins 0 (*RX*) and 1 (*TX*) while serial communication is in progress.

Other Serial commands

Serial.end()
In order to close the serial interface and to free the pins for other purposes, you write `Serial.end()`. Now the pins 0 and 1 can once again be used as digital inputs and outputs.

Serial.read()
The function `Serial.read()` reads a single byte, for instance one that has been sent from another Arduino board by means of `Serial.write(var)`. You can also send single bytes from a PC. This is possible with the terminal program *HTerm*, among others. You can download it for free at *http:///www.der-hammer.info/terminal/*.

```
byte x;
x = Serial.read();
    // The received byte is passed to x
```

Serial.available()
`Serial.available()` tells you whether a character is available in the serial buffer or not. This function comes in handy when you want to skip a program block if there

are no data in the buffer. Why waste processing power by trying to process data when there are none or when they have not changed since the last pass?

This function is very important for handling the serial interface. The following small test program demonstrates how it works. The output in Terminal is the corresponding ASCII code for the character. Enter a character into the top line of Terminal and click *Send*. Subsequently, the ASCII code is shown in Terminal. When you try this with A, you will see the ASCII code 65.

Example: *Serial.available*

```
// Franzis Arduino
// Serial.available function

byte input;
byte output

void setup()
{
    Serial.begin(9600);
    Serial.println("Serial.available function");
    Serial.println();
}

void loop()
{
    if (Serial.available() > 0)
    {
        input=Serial.read();
        Serial.print("I received the
            following character: ");
        output=char(input);
        Serial.println(output);
    }
}
```

Time needed: 10 min
Difficulty: 2

■ ■ ☐ ☐ ☐

This example also demonstrates type conversions. If you did not convert the type, the character, instead of the corresponding ASCII code, would be printed. Try to replace `Serial.print(output)` with `Serial.print(input)`. Now you will see the character in Terminal, not the ASCII code.

Serial.flush()

How to clear the serial buffer

This function removes the content of the serial buffer. It is used to erase the buffer in a reliable way after assigning the serial data. When data become invalid because of a communication error, you flush the buffer and try again.

Serial.print()

Writing characters

`Serial.print()` writes the data inside the serial buffer. Possible data types are integers, bytes, characters and float numbers.

```
// Text output
Serial.print("Hello world")
```

```
// Writes the ASCII string "79"
int b = 79;
Serial.print(b);
```

```
// Serial.print(b, DEC) with the format
// specification DEC writes the number as
// an ASCII string
int b = 79;
Serial.print(b, DEC);
```

```
// Serial.print(b, HEX) with the format
// specification HEX writes the number in the UART
// buffer as an ASCII string in hexadecimal format
int b = 79;
Serial.print(b, HEX);
```

```
// Serial.print(b, OCT) with the format
// specification OCT writes the number in the UART
// buffer as an ASCII string in octal format
int b = 79;
Serial.print(b, OCT);
```

```
// Serial.print(b, BIN) with the format
// specification BIN writes the number in the UART
// buffer as an ASCII string in binary format
int b = 79;
Serial.print(b, BIN);
```

The following example writes an ASCII table to the Terminal window. The enumeration begins with the value 33 because the preceding numbers are nonprintable control characters.

Example: Serial.print

```
// Franzis Arduino
// Serial.print function

void setup()
{
    Serial.begin(9600);
    Serial.println("ASCII table");
}

void loop()
{
    Serial.write(cnt);
    Serial.print(", dec: ");
    Serial.print(cnt);
    Serial.print(", hex: ");
    Serial.print(cnt, HEX);
    Serial.print(", oct: ");
    Serial.print(cnt, OCT);
```

Time needed: 10 min
Difficulty: 2

■ ■ □ □ □

```
Serial.print(", bin: ");
Serial.println(cnt, BIN);
if(cnt==126)while(1);
cnt++;
}
```

Serial.println()
Serial.println() writes the data inside the serial port, followed by an automatic line break in the form of a carriage return and a line feed. This command uses the same syntax as Serial.print().

Serial.write()
Serial.write() writes the data inside the serial port as bytes in binary form. There are several variants:

- Serial.write(*val*)
- Serial.write(*str*)
- Serial.write(*buf, len*)

The parameters have the following meaning:
- *val:* Sends a single byte.
- *str:* Sends a string byte by byte
- *buf:* Sends an array byte by byte
- *len:* Length of the array (buffer)

Example: Serial.print

Time needed: 10 min
Difficulty: 2
■ ■ ☐ ☐ ☐

```
// Franzis Arduino
// Serial.write function

byte val = 65;
char str[] = "Test";
byte buf[] = {'H','e','l','l','o'};
byte len = 3;
```

```
void setup()
{
   Serial.begin(9600);
   Serial.println("Serial.write function");
   Serial.println();
}

void loop()
{
   Serial.println("ASCII characters");
   Serial.write(val);
   Serial.println();

   Serial.println("String 1");
   Serial.write(str);
   Serial.println();

   Serial.println("String 2");
   Serial.write(buf, len);
   Serial.println();

   while(1);
}
```

The parameter `len` of `Serial.write(buf, len)` is specified as 3. Thus only the first three characters of `buf[]` are printed, which results in the output *Hel*.

Reading characters via the serial interface

So far you have learned how to write characters and strings and how to receive single characters. Furthermore, you have gained some basic knowledge of interacting with the Arduino and of entering words. In order to receive complete strings you have to write a little program that uses existing functions to receive whole words and sentences.

Receiving complete strings

The program waits until there are characters in the buffer and reads them into the *string array*. When more than 20 characters are received or a CR is read, the array is printed to Terminal. To make this possible, you have to change the default value *no line break* in the lower combo box of Terminal to *CR*. When you now enter *Hello* and click *Send*, the whole string is output in Terminal. Our Arduino program reads the received characters and sends them back to Terminal (echo).

Example: Serial.print

Time needed: 10 min
Difficulty: 2

```
// Franzis Arduino
// Serial.read

#define INLENGTH 20
#define INTERMINATOR 13

char inString[INLENGTH+1];
int inCount;

void setup()
{
   Serial.begin(9600);
   Serial.println("Serial.read");
   Serial.println();
   Serial.println("Enter a text of
       max. 20 characters: ");
}

void loop()
{
   inCount = 0;
```

```
do
{
    while (Serial.available()==0);
    inString[inCount] = Serial.read();
    if(inString[inCount]==INTERMINATOR) break;
}while(++inCount < INLENGTH);

inString[inCount] = 0;
Serial.print(inString);
}
```

Serial output with calculations

The following example shows a useful little program for the serial interface that converts temperature values from Celsius to Fahrenheit and vice versa. It also demonstrates the usage of functions. Furthermore, you can see that it is possible to write lines of code like the following:

A useful conversion program

```
res = grad * 9; res = res/5; res = res + 32;
```

However, you should use this notation only when it does not obscure the program flow.

Example: Celsius/Fahrenheit converter

```
// Franzis Arduino
// Celsius-Fahrenheit converter

float Celsius = 25.5;
float Fahrenheit = 88.2;

void setup()
{
    Serial.begin(9600);
    Serial.println("Celsius-Fahrenheit converter");
```

Time needed: 5 min
Difficulty: 2
■ ■ □ □ □

```
    Serial.println();
}

void loop()
{
    Serial.print(Celsius);

    Serial.print("Celsius temperature is");

    Serial.print(Celsius_to_Fahrenheit(Celsius));

    Serial.println("Fahrenheit");
    Serial.println();

    Serial.print(Fahrenheit);
    Serial.print("Fahrenheit temperature is");
    Serial.print(Fahrenheit_to_Celsius(Fahrenheit));
    Serial.println("Celsius");
    Serial.println();
    while(1);
}
float Celsius_to_Fahrenheit(float Celsius)
{
    float res;
    res = Celsius * 9 ; res = res / 5 ;
        res = res + 32;
    return res;
}


float Fahrenheit_to_Celsius(float Fahrenheit)
{
    float res;
    res = Fahrenheit - 32 ; res = res * 5 ;
        res = res / 9;
    return res;
}
```

Software UART

When you want to operate several serial devices with a UART interface on a microcontroller with only one available hardware UART, you have the option of emulating a UART interface in the software. Software UART uses the digital pins 0 and 1. Data are read into a 64-byte circular buffer. The drawback of software UART is the need for significant system resources. Let us take a look at the limitations of software UART:

A UART interface can be emulated in the software

- Transfer rate max. 9600 baud
- `Serial.read()` waits until there are data in the buffer
- `Serial.read()` must be executed in a loop when the function is not called, and incoming data get lost.

Tip More information about circular buffers can be found at:

http://en.wikipedia.org/wiki/Queue_%28abstract_data_type%29

The software UART library of the Arduino IDE offers the following functions:

Functions in the UART library

- `SoftwareSerial()`
- `begin()`
- `read()`
- `print()`
- `println()`

The configuration for the software UART may look as follows:

Example: Software UART

Time needed: 10 min
Difficulty: 2
■ ■ ☐ ☐ ☐

```
// Franzis Arduino
// Software UART

// Include the software UART library
#include <SoftwareSerial.h>

#define rxPin 2
#define txPin 3
#define ledPin 13

// Configure the software UART
SoftwareSerial mySerial = SoftwareSerial(rxPin,
    txPin);
byte pinState = 0;

void setup()
{
    // Configure the pins
    pinMode(rxPin, INPUT);
    pinMode(txPin, OUTPUT);
    pinMode(ledPin, OUTPUT);

    // Set the serial data transfer rate
    mySerial.begin(9600);
}

void loop()
{
    // Listen for incoming data
    char someChar = mySerial.read();

    // Print the received character
    mySerial.print(someChar);
```

```
    // Toggle the LED
    toggle(13);
}

void toggle(int pinNum)
{
    digitalWrite(pinNum, pinState);
    pinState = !pinState;
}
```

Thanks to the library that does much of the work for us, the program remains very clear and readable. The library is included by #include<SoftwareSerial.h>. A library contains the required functions for an extension. Basically, it consists of function calls like those we have used so far. You can easily access functions like the ones for implementing a software UART by just including the appropriate library. The libraries can be found in the folder of the Arduino IDE under:

A library contains the required functions for an extension

\\arduino-1.04\libraries

When you navigate to the directory \SoftwareSerial and open the program code (*.cpp file) in an editor, you can see how much additional code is processed in our little program. I recommend the editor Notepad++ for opening these files. You can download it for free at:

http://notepad-plus-plus.org/

The libraries are not written in the programming language for the Arduino, but in C++. More information about Arduino libraries can be found at:

More information about Arduino libraries

http://playground.arduino.cc/Code/Library

9.4.17 | Digital inputs and outputs

The program must know which pins you want to use as inputs or outputs. Therefore you have to specify this at the beginning of the code in the `void setup()` routine as you did for the LED L in previous programs. The following examples will show other possibilities.

If you omit the pin configuration, all pins of the microcontroller will be defined as inputs with a high resistance.

pinMode(pin, mode)

Use pinMode to configure a pin as input or output

`pinMode(pin, mode)` is used in the `void setup()` routine to configure a given pin as input or output.

```
pinMode(pin, OUTPUT)    // Configures pin as output
```

The Arduino microcontroller comes with convenient 20 kΩ pull-up resistors that can be activated in the program code. This makes it possible to attach a push-button at a digital pin in such a way that the button switches against ground (GND) when it is pressed.

In the open state (button not pressed), a voltage of ca 5 V is applied to the digital pin by the pull-up resistor. In this way, the register in the microcontroller senses a high state (1). When the button is pressed, the 5 V voltage is set to ground by the internal pull-up resistor. The voltage at the digital pin drops to 0 V, and the register now holds the value 0.

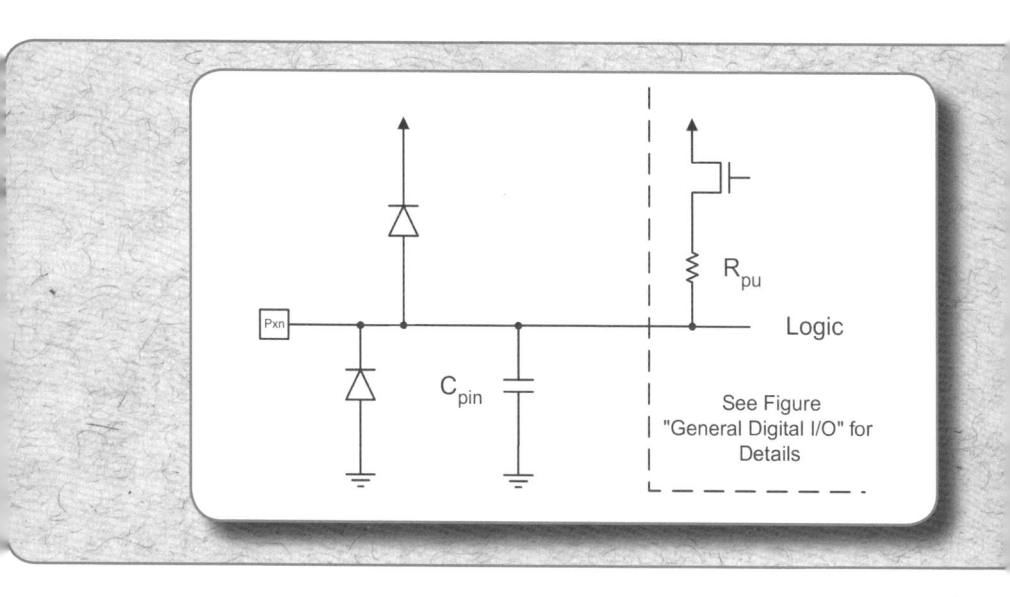

Figure 9.5: P_{xn} *is the physical pin on the Arduino board. The two diodes protect the port against static electricity. The capacitor* C_{pin} *provides the input capacitance for the pin. It amounts to just a few picofarad and is of no further significance for us.* R_{pu} *is the pull-up resistor. By means of the program code and a FET, it is switched against the operating voltage (5 V). When* P_{xn} *is set to ground, the logic contains the value 0 because no voltage is applied to it.*

You can access the internal pull-up resistors in the following way:

```
pinMode(pin, INPUT);
    // Configures pin as input
digitalWrite(pin, HIGH);
    // Activates the pull-up resistor
```

In the example above, the pin is not defined as output although it is written to. The write is just a method to activate the pull-up resistor. First, the digital pin has to be set as `INPUT` with `pinMode`. To activate the pull-up resistor you use `digitalWrite` to write `high` to the register of the pin. Now the pull-up resistor is active.

You can easily test this with a multimeter. Read the voltage with the positive probe at the pin and with the negative probe at ground (GND). The multimeter will show 5 V. When you remove the `digitalWrite` command, the multimeter will read 0 V because the resistor is no longer active.

With a button or switch, we can now apply 5 V to the pin. When the button is pressed, the 5 V are applied to the digital input. The register reads `high` (which is internally represented as 1) and signals that there is a voltage at the digital pin.

Possible errors

This method is rather error-prone because of the high resistance of the inputs. External interferences, e.g. by switching on a motor, may cause high voltages on the connection wires so that the digital input can detect a turned-on state only once in a while. To remedy this problem, you attach a 10 kΩ pull-down resistor to an input without pull-up. This resistor is connected in parallel from the digital input to ground. This prevents interferences from affecting the input.

9 Arduino Programming Basics   153

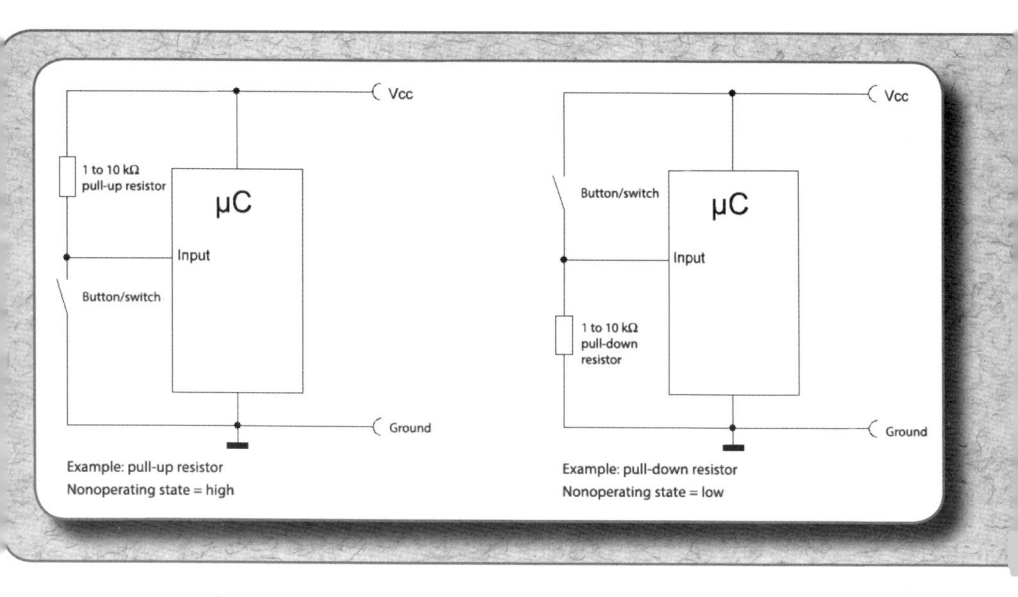

Figure 9.6: This diagram shows the circuit with an external pull-up and pull-down resistor, respectively. µC denotes the microcontroller, in our case the Arduino microcontroller on the Arduino board.

Pins configured as output are in a state of low impedance (low resistance), and any attached elements or circuits can only apply a maximum load of 40 mA per pin. This is enough to illuminate several LEDs (do not forget the serial resistor!), but not sufficient to operate relays, amps, magnetic coils, and motors.

Short circuits at the Arduino pins and excessive currents can destory the digital pins and even the whole Arduino board! In your experiments, you should therefore use a protective resistor for the output in order to limit the current.

Threats for the Arduino board

9.4.18 | Experiments with the Arduino

As you now have some knowledge about the configuration of digital pins, you may want to put it into practice.

First, you use the `digitalWrite()` statement to read a value at a digital pin. The return value is `high` or `low` which corresponds to 1 or 0, respectively. The pin can be specified as a variable or as a constant (0–13).

```
value = digitalRead(pin);
    // 1 or 0 is written to value
```

With `digitalWrite()` you set a digital output to `high` or `low` (1 or 0). The pin can be specified as a variable or as a constant (0–13).

```
digitalWrite(pin, value);
    // Switches the output
digitalWrite (pin, HIGH);
    // Sets pin to high (+5 V)
```

Reading the button state at the digital input

The following example reads the state of a push-button attached to the digital input pin 12 und forwards it to the LED. After the transfer, the LED L lights up immediately. This happens because we have activated the internal pull-up resistor, so that input pin 12 is `high`, that is 1.

When you now press the button, the input is switched to ground, so the LED is turned off.

Required parts for the experiment
- 1 x microcontroller board Arduino Uno
- 1 x breadboard
- 1 x push-button
- 2 x hook-up wire, ca. 5 cm

9 Arduino Programming Basics | 155

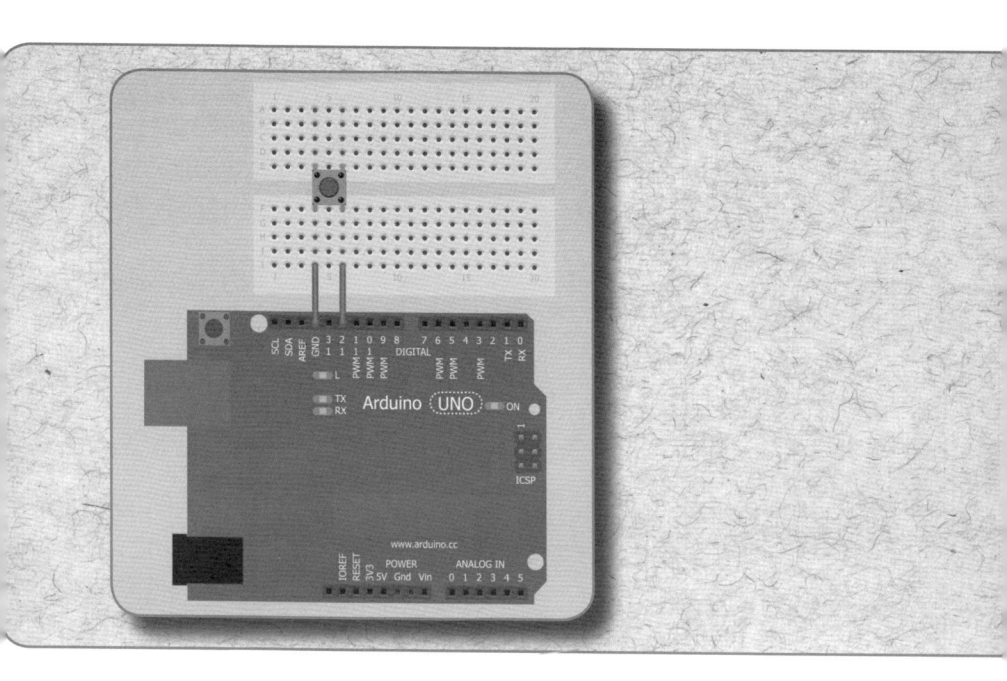

Figure 9.7: Set-up of the circuit for the experiment

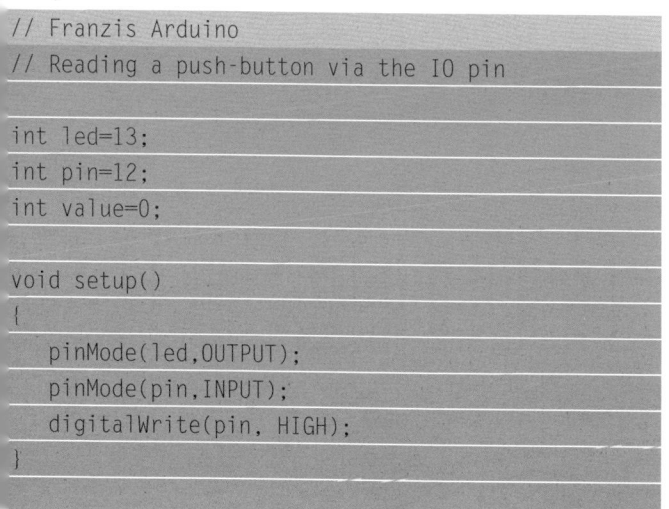

Example: IO

```
// Franzis Arduino
// Reading a push-button via the IO pin

int led=13;
int pin=12;
int value=0;

void setup()
{
    pinMode(led,OUTPUT);
    pinMode(pin,INPUT);
    digitalWrite(pin, HIGH);
}
```

Time needed: 10 min
Difficulty: 2

■ ■ □ □ □

```
void loop()
{
    value=digitalRead(pin);
    digitalWrite(led,value);
}
```

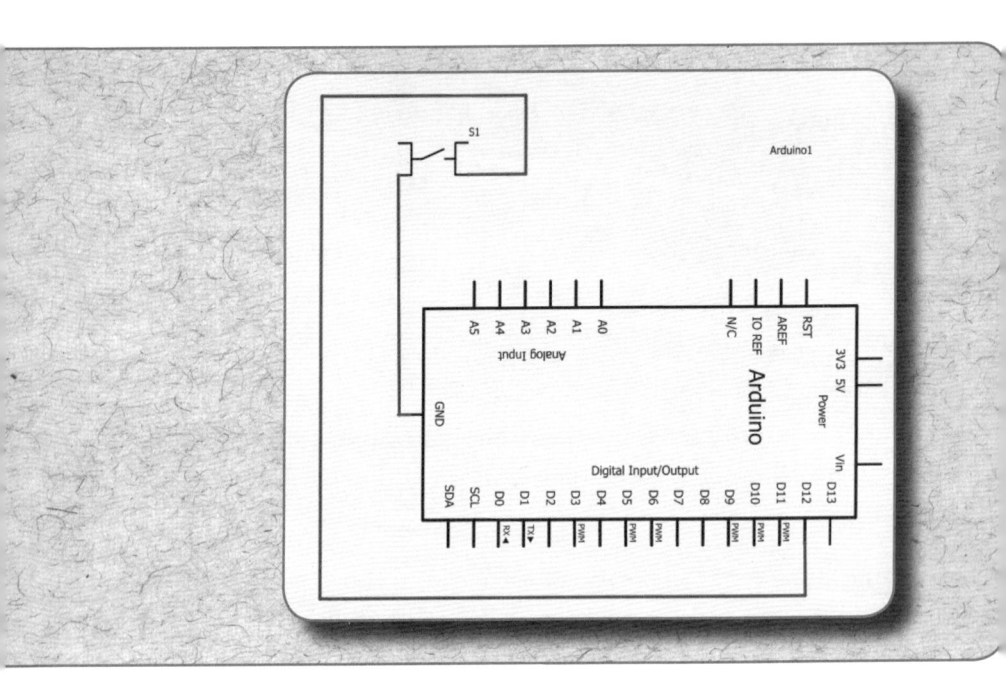

Figure 9.8: The circuit diagram for the experiment. S1 is the push-button.

The total load on the microcontroller must not exceed 200 mA!

Info A pin of the ATmega can handle up to 40 mA. The total load on the microcontroller must not exceed 200 mA (depending on the type of housing). There are also differences between the individual pins. To be sure, you can always check the data sheet.

Push-button with external pull-down resistor

In the previous example, the button was operated by means of the internal pull-up resistor. The following example demonstrates how to read the button state using a pull-down resistor. LED L lights up only when we apply +5 V to the digital input 12.

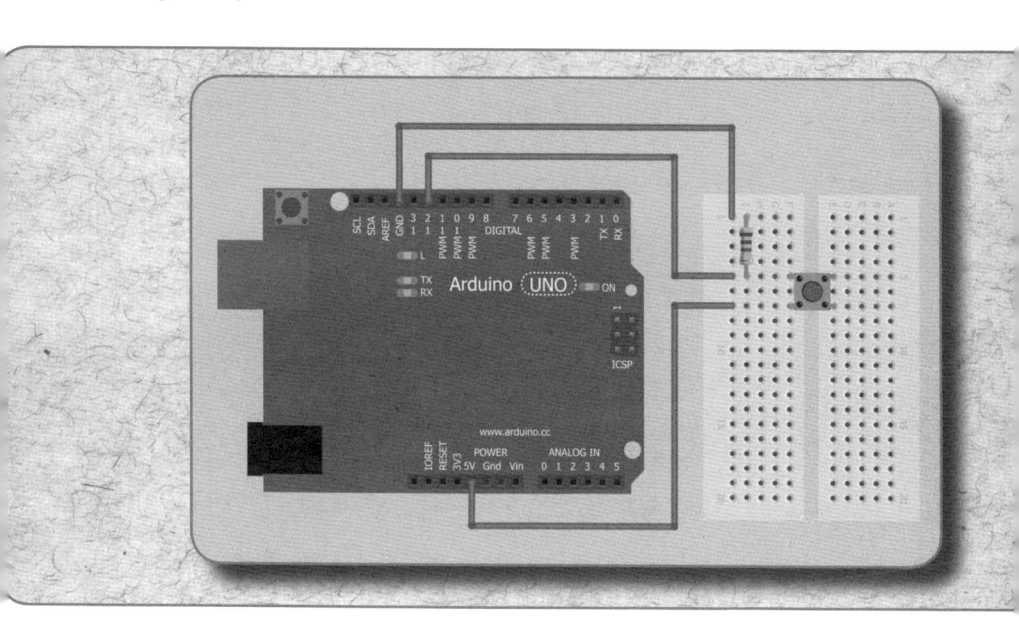

Figure 9.9: Set-up of the circuit with pull-down resistor

- 1 x microcontroller board Arduino Uno
- 1 x breadboard
- 1 x push-button
- 1 x 10 kΩ resistor
- 3 x hook-up wires, ca. 10 cm

Required parts for the experiment

Time needed: 10 min
Difficulty: 2
■ ■ ◻ ◻ ◻

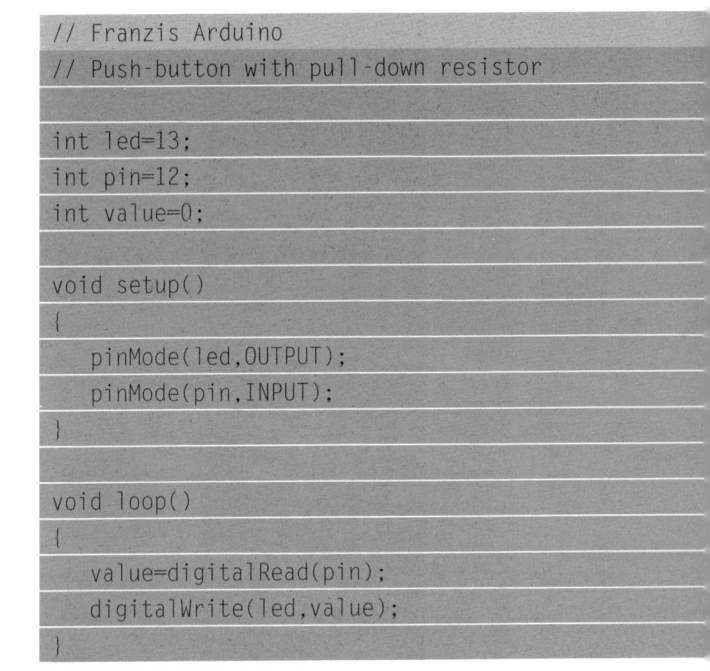

Example: Pull-down

```
// Franzis Arduino
// Push-button with pull-down resistor

int led=13;
int pin=12;
int value=0;

void setup()
{
    pinMode(led,OUTPUT);
    pinMode(pin,INPUT);
}

void loop()
{
    value=digitalRead(pin);
    digitalWrite(led,value);
}
```

Push-button with external pull-up resistor

Circuit for using an external pull-up resistor with the Arduino

This example shows how to set up a circuit with the Arduino and an external pull-up resistor. This makes sense when you want to use the circuit in an environment with heavy interference (e.g. by motors) and the resistance of the internal pull-up resistor is too high. Especially with long wires (more than 5 m), an external pull-up resistor with a small resistance value (less than 5 kΩ) is recommended.

Here, the resistor is not attached to ground but to the operating voltage of +5 V. The LED lights up at power-on and is turned out when you press the button (as it was in the set-up with the internal pull-up resistor).

9 Arduino Programming Basics

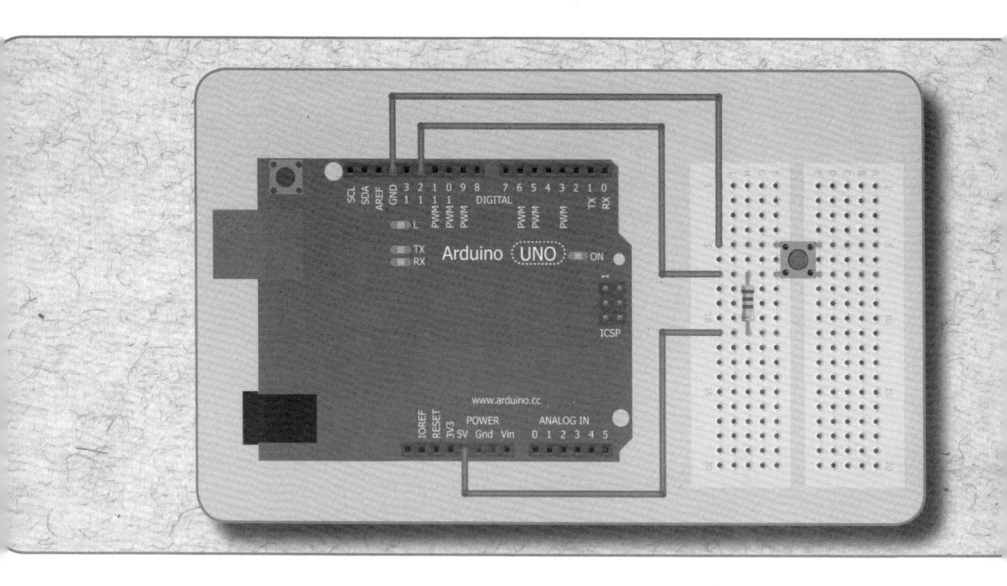

Figure 9.10: Set-up of the circuit with an external pull-up resistor

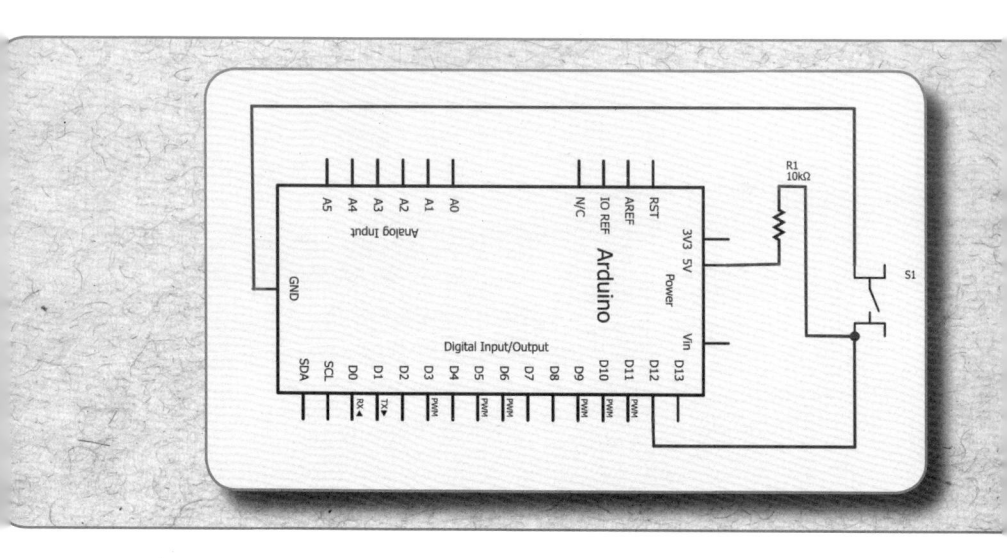

Figure 9.11: Circuit diagram with external pull-up resistor

| Required parts for the experiment | ■ 1 x microcontroller board Arduino Uno
■ 1 x breadboard
■ 1 x push-button
■ 1 x 10 kΩ resistor
■ 3 x hook-up wires, ca. 10 cm |

Example: *External pull-up resistor*

Time needed: 10 min
Difficulty: 2

```
// Franzis Arduino
// Push-button with external pull-up resistor

int led=13;
int pin=12;
int value=0;

void setup()
{
    pinMode(led,OUTPUT);
    pinMode(pin,INPUT);
}

void loop()
{
    value=digitalRead(pin);
    digitalWrite(led,value);
}
```

9.4.19 | ADC analog input

The Arduino board comes with an internal analog/digital converter (ADC) so that you can use it to read voltages. The ADC used in the ATmega has a resolution of 10 bit, i.e. with a reference voltage of 5 V, one digit represents 0.0048 V = 4.88 mV.

The internal analog/digital converter

The microcontroller board has six analog inputs – 0 to 5. They are fed into a *multiplexer* in the microcontroller. The multiplexer connects the input specified in the program code with the ADC. The microcontroller itself has only one ADC. The inputs are switched to the multiplexer, and then the voltage is measured. The voltage applied to the analog input is then converted into a digital value from 0 to 1023. This value can be processed in the program code. Each digital value (or *step*) represents 0.0048 V = 4.88 mV.

You can calculate the resolution with the following simple formula:

```
step = V_ref / resolution 2ⁿ bit
     = 5 V / 1024
     = 0.0048 V = 0.88 mV
```

Calculating the resolution

The raw digital value can be calculated as follows:

$$\text{raw value} = \frac{(1023 \times \text{applied voltage [ADC]})}{V_{ref}}$$

The precision varies between ±2 steps. With a reference voltage of 5 V, we thus have a precision of ±0.0097 V. The ADC measures the applied voltage accurately to two decimal places. This requires of course a stable reference voltage!

analogRead(pin)

`analogRead(pin)` reads the value of the specified pin with a resolution of 10 bit. This function is only available for the analog inputs 0 to 5. The resulting integer value have a range from 0 to 1023.

```
value = analogRead(pin)
        // The digital value of the ADC is assigned
        // to value
```

Info As opposed to their digital counterparts, the analog pins do not have to be configured as input or output. They can be read immediately with `analogRead()`.

Practice program: voltage reading

To close this topic, we will look at a little program that reads the voltage at ADC input 0.

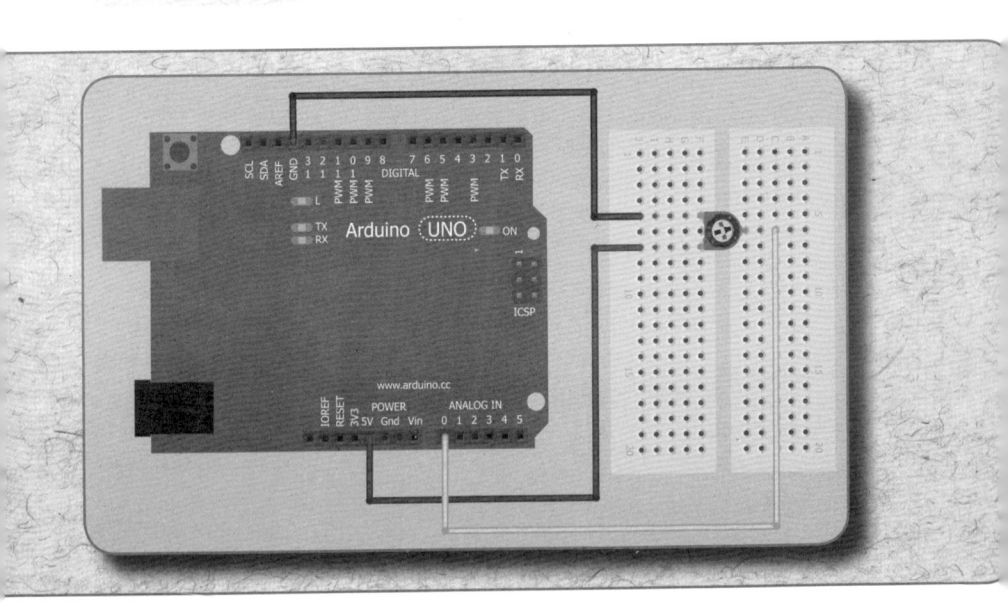

Figure 9.12: Set-up of the circuit with pull-down resistor

9 Arduino Programming Basics

- 1 x microcontroller board Arduino Uno
- 1 x breadboard
- 1 x 10 kΩ trimmer
- 3 x hook-up wire, ca. 10 cm

Required parts for the experiment

Example: ADC

```
// Franzis Arduino
// ADC

int ADC0=0;
int value;
int LEDpin=13;

void setup()
{
    Serial.begin(9600);
}

void loop()
{
    value=analogRead(ADC0);
    Serial.print("ADC0 = ");
    Serial.println(value);
    delay(1000);
}
```

Time needed: 10 min
Difficulty: 2

■ ■ □ □ □

When you turn the potentiometer (the trimmer), you will see the digital value changing. Depending an the position of the potentiometer knob the output shows a value from 0 to 1023. It is possible that you cannot fully reach 1023, because some potentiometers have a residual resistance even if you turn it to maximum. Because of this,

Different readings indoors and outdoors

you always lose a little voltage so that you cannot reach the maximum value. When you leave open the ADC or when you use a longer wire, the readings will vary heavily. The reason for this is the high resistance of the ADC input (ca. 100 kΩ). By means of the wire that acts as an antenna, the ADC can now receive the electromagnetic fields in the environment. When you draw closer to an electrical device, the value will increase significantly. Outdoors, the value is relatively small, provided you do not stand next to the pylon of a high-voltage power line. When you damp the input with an additional resistor between the analog input and ground (say, 100 kΩ), the input gets less sensitive, so that you can detect stronger fields.

The following experiment can be carried out with the existing set-up. This example is an extension to the blinking light you encountered as a test program at the beginning of this book. With the ADC extension, you can control the blink frequency of the LED by the trimmer. To this end, you write the ADC value (0 to 1023) into the variable for the delay. As the delay has to be specified in milliseconds, you can set it to 0 to 1023 ms by means of the trimmer.

Extended blinking light

Leave the ADC input open or attach a small length of wire to it. This way, the blinking frequency varies randomly, because the ADC value is changed by electromagnetic fields in the environment. In order to affect the blinking frequency, it is sufficient to approach the short wire with your hand.

Example: ADC blinker

```
// Franzi's Arduino
// Controlling the LED blinking frequency
// via the ADC

int ADC0=0;
int value;
int LEDpin=13;

void setup()
{
    pinMode(LEDpin,OUTPUT);
}

void loop()
{
    value=analogRead(ADC0);
    digitalWrite(LEDpin, HIGH);
    delay(value);
    digitalWrite(LEDpin, LOW);
    delay(value);
}
```

Time needed: 10 min
Difficulty: 2

■ ■ □ □ □

4.20 | Analog PWM outputs

Six PWM outputs are available on the Arduino Uno: pins 3, 5, 6, 9, 10, and 11. The older ATmega8 controllers provide only the pins 9, 10, and 11. The PWM outputs can be used for D/A conversion, for controlling DC motors and servos, and to output audio frequencies. Pulse width modulation (PWM) creates a digital output signal with a modulated duty cycle. The *duty cycle* determines the ratio of the length of the turned-on state to the oscillation period. With different duty cycles, the frequency and level of the signal stay the same. Only the ratio of the length of the *high* to the *low* level changes.

PWM creates a digital output signal with modulated duty cycle

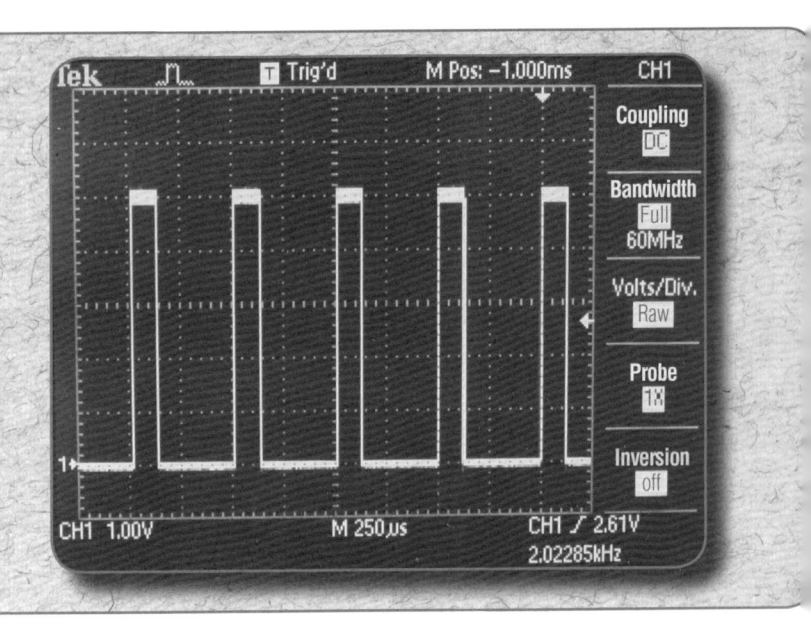

Figure 9.13: PWM with a duty cycle of 25%

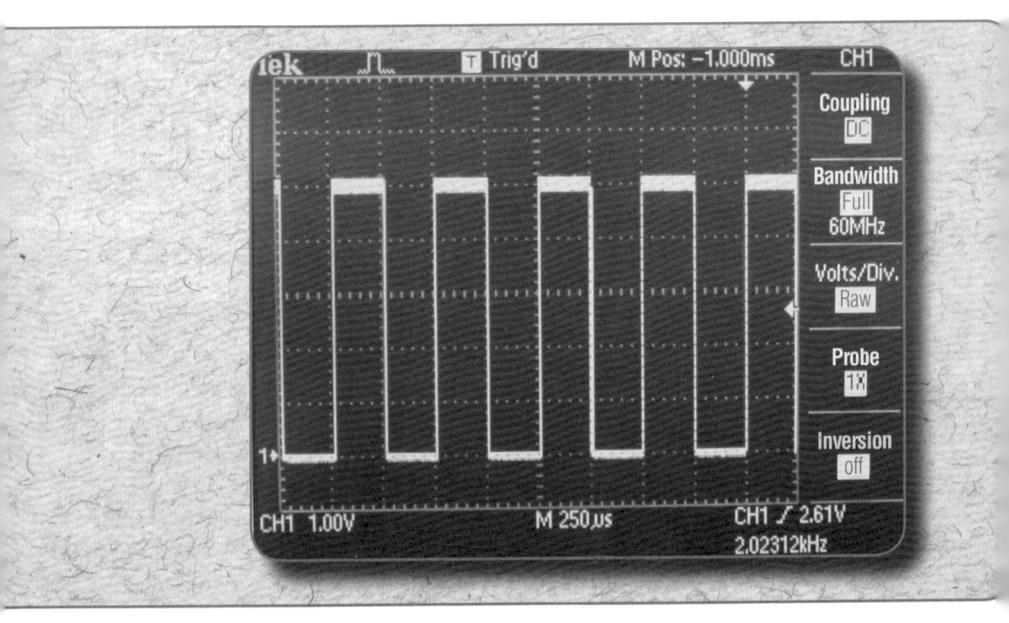

Figure 9.14: PWM with a duty cycle of 50%

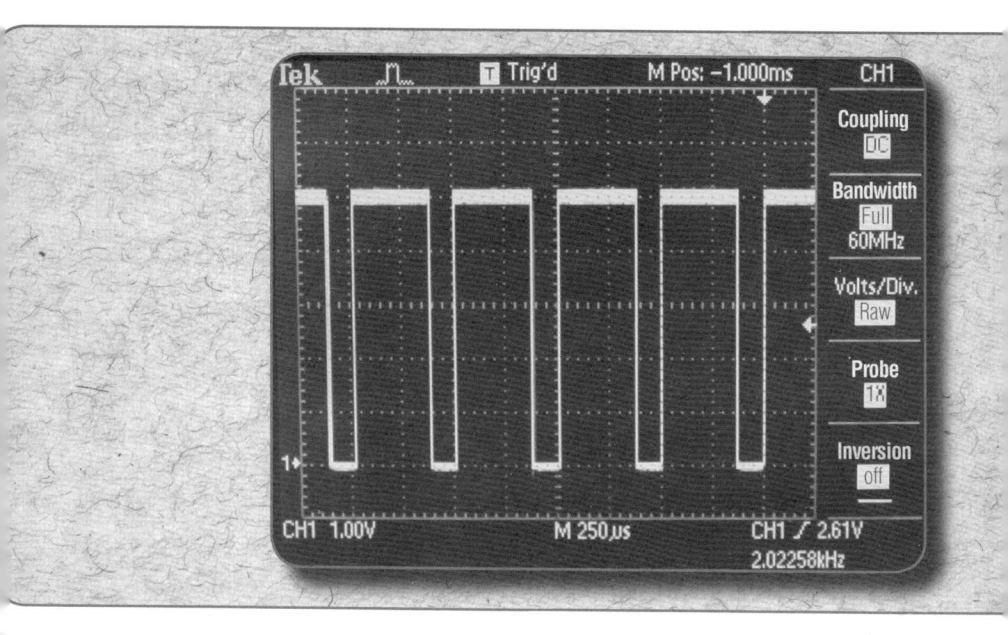

Figure 9.15: PWM with a duty cycle of 75%

analogWrite(pin, value)

This command changes the duty cycle at a PWM output pin. The value can be set to a variable or a constant in the range from 0 to 255.

```
analogWrite(pin, value)
    // The value can be a number from 0 to 255
```

The values 0 und 255 cause a consistent voltage of 0 or 5 V, respectively, at the specified analog output pin. With values greater than 0 and less than 255, the voltage at the pin changes rapidly between 0 V (*low*) and 5 V (*high*). The greater the value, the longer the pin is set to *high* (5 V). At 64, the pin has 0 V for three-quarters of the time and 5 V for one quarter.

With a value of 127, the output voltage is *high* for ha the time and *low* for the other half. Thus we have a dut cycle of 50%. When you use a value of 192, the pin is *lo* for one quarter of the time and *high* for the other three quarters. Since this function is hardware-based, the PWM generation runs independently of the program until yo change the duty cycle again with `analogWrite`. So yo set the PWM by `analogWrite`, and the hardware of th microcontroller generates the PWM signal without pu ting a load on the processor. The command only re-write the register and is fed to the PWM module.

Analog pins do not have to be configured as input or output

Info As opposed to their digital counterparts, the analo pins do not have to be configured as input or output.

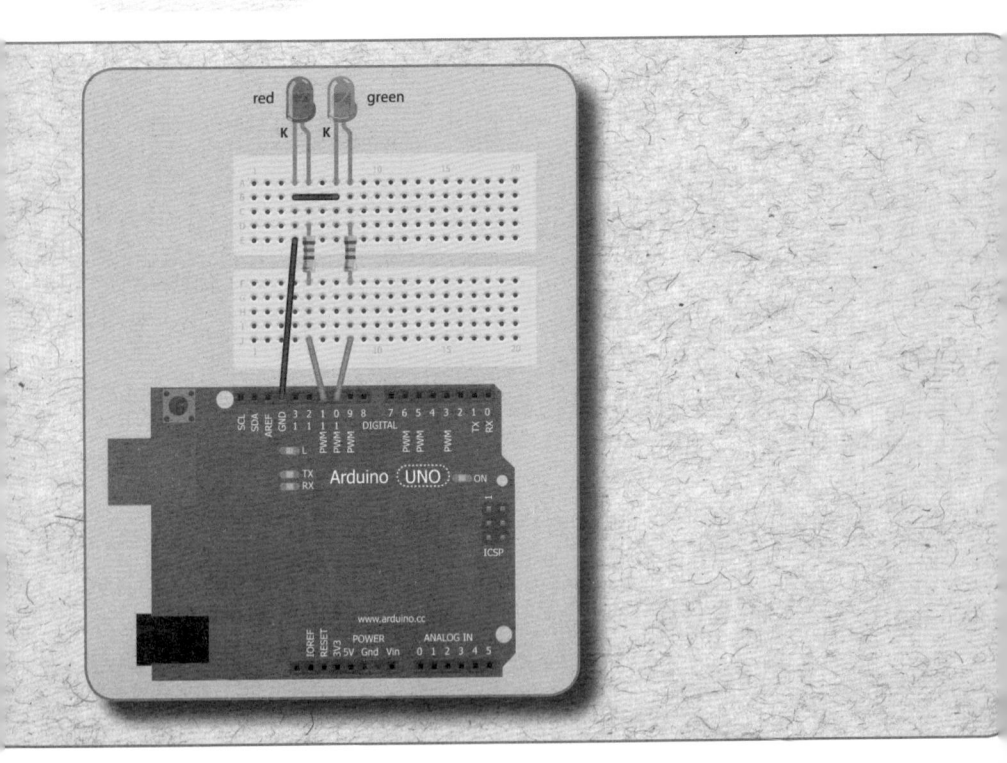

Figure 9.16: Set-up of the PWM circuit

9 Arduino Programming Basics

This sample project causes the LED to blink very often. The brightness is set by PWM.

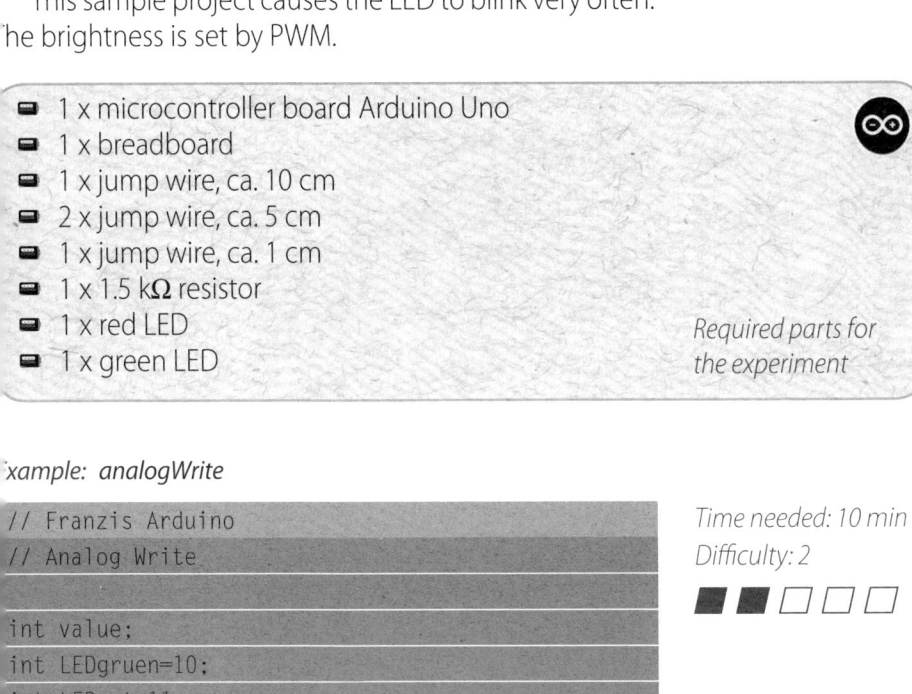

- 1 x microcontroller board Arduino Uno
- 1 x breadboard
- 1 x jump wire, ca. 10 cm
- 2 x jump wire, ca. 5 cm
- 1 x jump wire, ca. 1 cm
- 1 x 1.5 kΩ resistor
- 1 x red LED
- 1 x green LED

Required parts for the experiment

Example: analogWrite

```
// Franzis Arduino
// Analog Write

int value;
int LEDgruen=10;
int LEDrot=11;

void setup()
{
    // This time setup() contains nothing.
    // The hardware carries out the PWM work.
}

void loop()
{
    for(value=0;value<255;value++)
    {
        analogWrite(LEDgruen, value);
        analogWrite(LEDrot, 255-value);
        delay(5);
    }
}
```

Time needed: 10 min
Difficulty: 2

■ ■ ☐ ☐ ☐

```
    delay(1000);

    for(value=255;value!=0;value--)
    {
      analogWrite(LEDgruen, value);
      analogWrite(LEDrot, 255-value);
      delay(5);
    }

    delay(1000);
}
```

When you attach the small piezo buzzer to the analog output, you can make the pulses audible. However, the buzzer is not very loud.

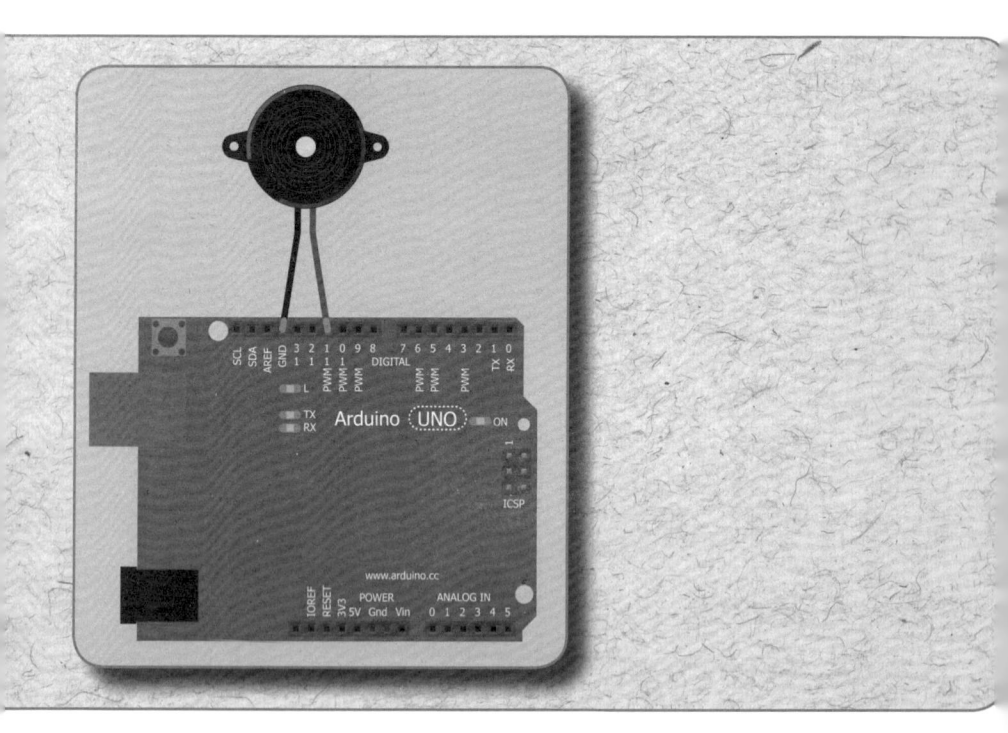

Figure 9.17: Set-up of the circuit to make the PWM audible

9.4.21 | Have a break, have a delay

In the previous examples we have used `delay()` to provide for a short break. This statement is not very complicated, but nonetheless we will give a more detailed explanation. `delay` simply halts the program for the specified number of milliseconds.

How does »delay« work?

delay(ms)
The program pauses for the specified number of milliseconds (1000 milliseconds = 1 second).

```
delay(1000);    // Waits for one second
```

micros(ms)
The program pauses for the specified number of microseconds (1000 microseconds = 1 millisecond).

```
micros(1000);   // Waits for one millisecond
```

9.4.22 | Random numbers

In writing measuring, control, or game programs, it is often useful to generate random numbers, for instance, when you want to turn on and off the lights in a house at different times to simulate the presence of the residents. For this purpose, you can use the `random` function.

Random numbers are used for simulation programs, among other things

randomSeed(seed)
This statement sets the initial value for the `random()` function. When you change the seed value, different random numbers are generated. If the seed value is constant at every program start or every call to `randomSeed`, then you always get the same random numbers. To avoid this,

it is usually helpful to read a seed value at the analog input that always shows a different state. Since the analog measurement always varies by some digits (or steps), you can even resort to an analog input where a fixed voltage is applied. However, it is better to use an analog pin with a great variance in the measurement results, like an open input with a piece of wire for an antenna.

```
randomSeed(value);
    // value sets the initial value
    // for the random function
```

random(min, max)

The `random` function generates pseudo-random numbers in the range between the specified minimum and maximum.

```
value = random(100, 200);
// Generates random numbers between 100 and 200
```

Info Use `random(min, max)` after `randomSeed()`. The `randomSeed` function sets the initial value for `random`.

Example: Random numbers

Time needed: 10 min
Difficulty: 2

```
// Franzis Arduino
// Random numbers

int x,y=0;

void setup()
{
    randomSeed(analogRead(0));
    Serial.begin(9600);
```

```
    Serial.println("Random numbers");
    Serial.println();
}

void loop()
{
    for(x=0;x<20;x++)
    {
        y=random(0, 10);
        Serial.print(y);
        Serial.print(",");
    }
    Serial.println();

    for(x=0;x<20;x++)
    {
        y=random(10,100);
        Serial.print(y);
        Serial.print(",");
    }
    Serial.println();

    for(x=0;x<20;x++)
    {
        y=random(0,x+1);
        Serial.print(y);
        Serial.print(",");
    }
    Serial.println();

    while(1);
}
```

In a test run, I got the following random distribution:

0, 9, 5, 5, 9, 3, 2, 1, 1, 9, 4, 3, 9, 9, 5, 6, 1, 0, 4, 8, 83, 24, 24
99, 92, 36, 97, 35, 13, 10, 43, 98, 88, 52, 89, 86, 29, 35, 37, 58
0, 0, 1, 3, 0, 1, 1, 4, 6, 9, 7, 3, 1, 3, 5, 8, 9, 9, 17, 18

A game of dice based on random numbers

You got different numbers? Well, no wonder! After al the numbers have to be random, because they are initial ized with the value at the analog input 0. Run the pro gram several times and make sure that it always print different numbers. All of these numbers are between *mir* and *max*.

Example: Game of dice

Time needed: 10 min
Difficulty: 2

```
// Franzis Arduino
// Game of dice

int i,number=0;
int Nmb=6;
int adc;

void setup()
{
    Serial.begin(9600);
    randomSeed(analogRead(0));
}

void loop()
{
    Serial.print("You threw the following
        numbers: ");
```

```
for(i=0;i<Nmb;i++)
{
    number=random(1,6);
    Serial.print(number);
    Serial.print(" ");
}

Serial.println();
Serial.println("Press reset to roll
    the dice again ...");
Serial.println();
while(1);
}
```

9.4.23 | A stopwatch program: How much time passed?

The Arduino IDE provides several specialized functions to measure the elapsed time since program start or since the execution of a function. Among other things, you can use these functions to create waiting loops that do not halt the whole program and wait for a given time like `delay()` does. The elapsed time can be determined in milliseconds and microseconds. The time measurement can also be used very effectively to control the program execution and optimize its speed.

A useful stopwatch program

millis()

`millis()` returns the elapsed time since the last call to this function. The value is given in milliseconds.

The value is stored in an `unsigned long` variable that is subject to an overflow after 50 days so that it starts again at 0. It is therefore not possible to retain the measurement longer than 50 days. However, this should be enough for most applications.

```
value = millis();
     // Returns the elapsed time in milliseconds
```

The sample program measures the run time of `loop()` and returns the result in Terminal.

Example: millis()

Time needed: 5 min
Difficulty: 1

```
// Franzis Arduino
// Time measurement 1

unsigned long value;

void setup()
{
  Serial.begin(9600);
  Serial.println("Arduino Time Measurement 1");
  Serial.println();
}

void loop()
{
  Serial.print("Time: ");
  value=millis();
  Serial.println(value);
  delay(1000);
}
```

Info 1000 milliseconds equal 1,000,000 microseconds.

micros()

micros() returns the time that has elapsed since the program was started. The value is given in milliseconds. On 16 MHz boards like the Arduino Uno, an overflow occurs after 70 minutes, so that the measurement is reset to 0.

```
value = micros():
    // Returns the elapsed time in microseconds
```

Again the sample program measures the run time of `loop()`. If you remove the `delay()` statement, you can see how much time the rest of the program code needs for passing the loop. When you insert several other `Serial.println` statements, the run time will increase significantly.

Example: micros()

Time needed: 5 min
Difficulty: 1

■ □ □ □ □

```
// Franzis Arduino
// Time measurement 2

unsigned long value;

void setup()
{
    Serial.begin(9600);
}

void loop()
{
    Serial.print("Time: ");
    value=micros();
    Serial.println(value);
    delay(1000);
}
```

TURN ON YOUR CREATIVITY

FRANZIS
ARDUINO

MORE EXPERIMENTS WITH THE ARDUINO

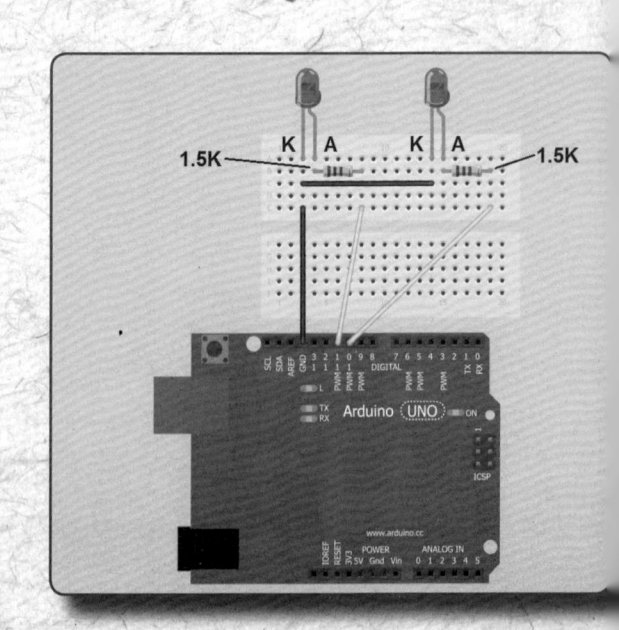

10

Even more experiments with the Arduino

Now that you have worked through the fundaments and made yourself familiar with programming the Arduino, you can start with hands-on experiments. The following projects build up on the basic knowledge you have gained in the previous chapter and extend it with new functions and programming options.

It is assumed that you already understand the program statements described so far, so you can implement the examples.

The basic mode of operation is given for all the examples, but there will be no further explanation of familiar statements. If you do not have a firm grasp on some the commands, you shall tackle them again.

Work with the breadboard

In most of the following experiments, you will need the breadboard and the components included in the tutorial kit. The circuits are deliberately kept simple. You can easily follow the current flow on the breadboard without a circuit diagram.

10.1 | LED dimmer

Build an LED dimmer for your living room

In the previous chapter, you have become acquainted with the analog PWM output and `analogWrite`. This allows you to build a dimmer that controls the brightness of an LED. Use a red LED at analog output 3 for the next experiment. If you want to use more powerful LEDs like those by Luxeon, you have to add a transistor to the analog output in order to increase the small current of the microcontroller to the amount needed by the LED.

The example project already uses a transistor as an amplifier and shows how to use it on a digital PWM output. In this experiment, we only use the low-current LED included in the tutorial kit, but you can apply a greater load to the collector circuit like the high-power LED mentioned above or a small lightbulb for a flashlight (max 100 mA). The push-buttons S1 (brighter) and S2 (darker) control the duty cycle of the PWM output and thus the brightness. The transistor relieves the digital pin. Only a very small current (ca. 300 times smaller than the load) flow to the base. This current is amplified by the transistor that uses the small base current to switch the larger collector current.

Required parts for the experiment

- 1 x microcontroller board Arduino Uno
- 1 x red LED
- 2 x push-buttons
- 1 x transistor BC548C
- 1 x 1.5 kΩ resistor
- 1 x 4.7 kΩ resistor
- 5 x jump wire, ca. 10 cm
- 2 x jump wire, ca. 5 cm

10 More Experiments with the Arduino 181

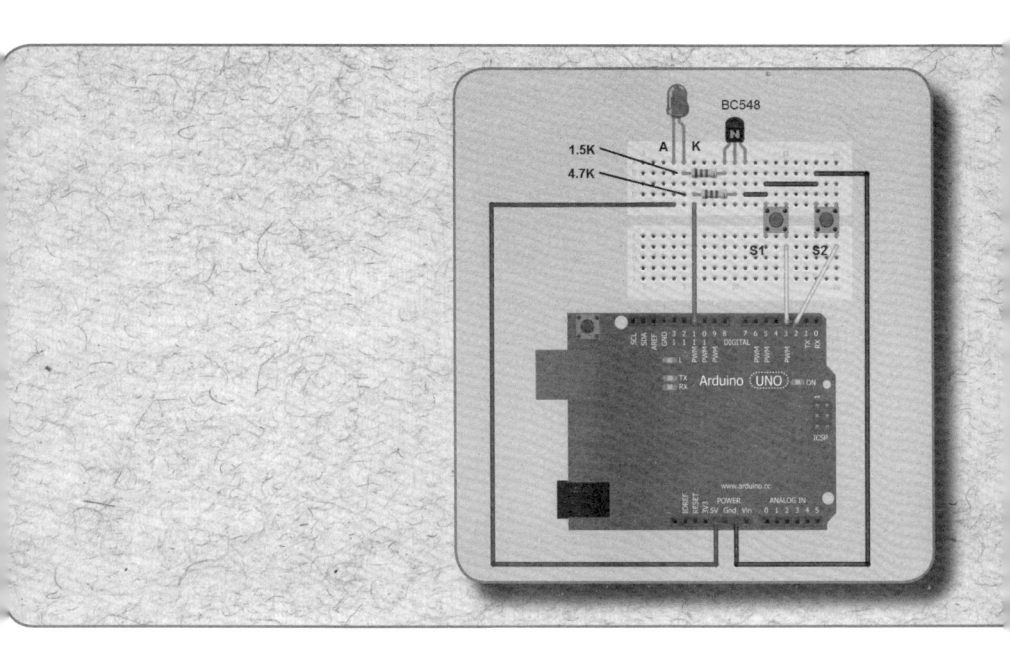

Figure 10.1: Diagram of the set-up for an LED dimmer with transistor

Example: LED dimmer

```
// Franzis Arduino
// LED-Dimmer

int brightness=0;
int SW1=3;
int SW2=2;
int LED=11;

void setup()
{
    pinMode(SW1,INPUT);
    digitalWrite(SW1,HIGH);
    pinMode(SW2,INPUT);
    digitalWrite(SW2,HIGH);
}
```

Time needed: 15 min
Difficulty: 2

■ ■ □ □ □

```
void loop()
{
   if(!digitalRead(SW1)&&digitalRead(SW2))
   {
      if(brightness<255)brightness++;
      analogWrite(LED,brightness);
      delay(10);
   }
   else if(digitalRead(SW1)&&!digitalRead(SW2))
   {
      if(brightness>0)brightness--;
      analogWrite(LED,brightness);
      delay(10);
   }
}
```

This example also demonstrates the usage of logical operators like ! and && in an if query. These comparison operations cause the push-buttons to lock each other, so that nothing happens when you press both of them simultaneously.

```
if(!digitalRead(SW1)&&digitalRead(SW2))
{
   // statement 1
}
else if(digitalRead(SW1)&&!digitalRead(SW2))
{
   // statement
}
```

The preceding program snippet can be verbalized as follows: If SW1 is *low* (0 V because the button is pressed and the pull-up resistor is active and we thus have a digi-

10 More Experiments with the Arduino

tal value of 0) and SW2 is *high* (the button is not pressed and the pull-up resistor is active, thus the digital value is 1), then execute the first block. If SW1 is *high* (not pressed, 5 V are applied, thus the digital value is 1) and SW2 is *low* (pressed, digital value is 0) then execute the code after `else if`.

In short form:

```
If SW1 = 0 and SW2 = 1 then execute statement 1
```

If the first condition is not true, then test the following:

```
If SW1 = 1 and SW2 = 0 then execute statement 2
```

In the statements we increment (++) or decrement (--) the variable `brightness`. To avoid an overflow, the less-than query (<) and the greater-than query (>) provide an upper limit of 255 and a lower one of 0.

No matter how long you press the button, the value of the variable will never exceed 255 or drop below 0. In order to provide a more convenient way to set the brightness, a delay of 10 ms is added. Every pass takes 10 ms, which makes the adjustment of the brightness very comfortable and simple.

A delay makes for a more pleasant dimming experience

When you increase the delay value, the dimming process will be slower when you press a button. If you remove `delay` completely, the variable `brightness` is incremented or decremented so rapidly that you cannot observe any dimming. Instead, it looks as if the LED was turned on or off.

10.2 | Soft flasher

More lighting effects by using the sine function

With the sine function, you can coax the analog output to issue a sinusoidal signal. This provides for a smooth increase and decrease in the brightness of the LED which comes in handy for some applications. This slow variation in the brightness looks as if the board had a beating heart.

The set-up is the same as in the previous example (see Fig. 10.1). The main program runs through a loop that counts from 0 to 255. The corresponding values to the numbers are retrieved from the array with the sine function table and passed as PWM values to the analog output via `analogWrite`. Using a table is significantly faster than calculating the values at run time.

Example: Sine wave blinker

Time needed: 10 min
Difficulty: 2
■ ■ □ □ □

```
// Franzis Arduino
// Sine wave blinker

byte i=0;
int LED=11;

byte Data[] = {128,131,134,137,140,144,147,150,153,
156,159,162,165,168,171,174,177,180,182,185,188,191,
194,196,199,201,204,206,209,211,214,216,218,220,222,
224,226,228,230,232,234,236,237,239,240,242,243,244,
246,247,248,249,250,251,251,252,253,253,254,254,254,
255,255,255,255,255,255,255,254,254,253,253,252,252,
251,250,249,248,247,246,245,244,242,241,240,238,236,
235,233,231,229,227,225,223,221,219,217,215,212,210,
208,205,203,200,197,195,192,189,187,184,181,178,175,
172,169,167,164,160,157,154,151,148,145,142,139,136,
133,130,126,123,120,117,114,111,108,105,102,99,96,
14,12,11,10,9,8,7,6,5,4,4,3,3,2,2,1,1,1,1,1,1,1,2,
```

```
2,2,3,3,4,5,5,6,7,8,9,10,12,13,14,16,17,19,20,22,
24,26,28,30,32,34,36,38,40,42,45,47,50,52,55,57,60,
62,65,68,71,74,76,79,82,85,88,91,94,97,100,103,106,
109,112,116,119,122,125,128};

void setup()
{
    // This time, we do not have to do anything
    // in here ...
}

void loop()
{
    for(i=0;i<255;i++)
    {
        analogWrite(LED,Data[i]);
        delay(5);
    }
}
```

The program uses the dynamic byte Array `Data[]` whose values are assigned in the braces. It is called a dynamic array because its size is determined by the number of values defined in the braces. As we have put 256 values into the braces, the array has a size of 256 bytes. You can access the single values in the array by the index in the range from 0 to 255.

The code fetches one value (`Data[i]`) at a time and writes it to the hardware using `analogWrite`, thereby changing the duty cycle of the PWM output.

How to change the duty cycle of the PWM output

By the way: This is only a quasi-analog output. The statement is called `analogWrite`, but we only change the duty cycle of the output. Without a filter at the output, we just get a simple PWM signal and not a true analog signal as it is issued by a real digital/analog converter (DAC). The next experiment will show how you can generate a real analog signal out of a PWM signal.

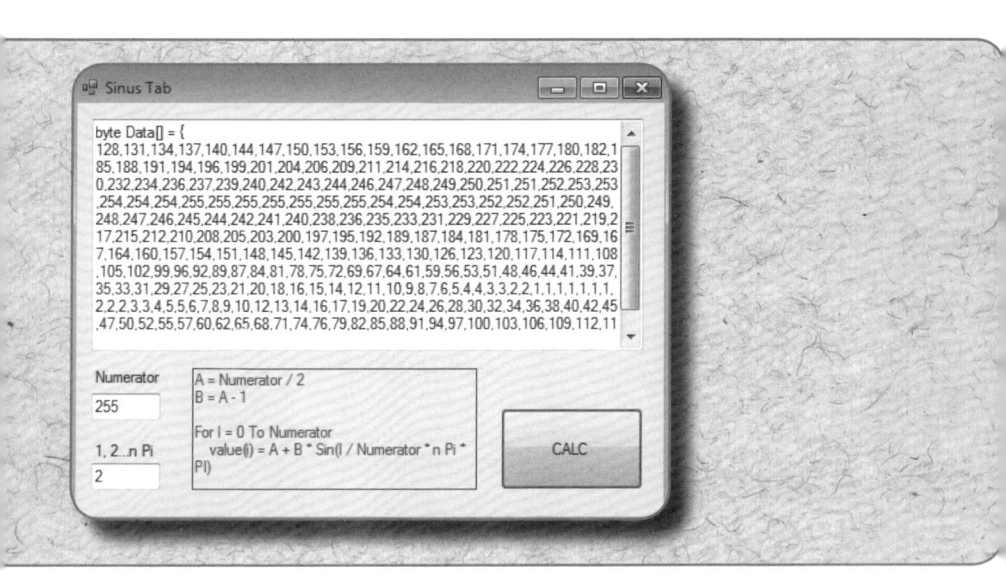

Figure 10.2: The program Sinus Tab that calculates the values of the sine wave table

The little Visual Basic .NET program Sinus Tab calculates the sine wave table that you can directly insert into your programs. You can find the program on the CD included in the tutorial kit.

Variants with an oscilloscope

If you happen to own an oscilloscope you can attach an RC circuit to the analog output (instead of the LED) and view the sinusoidal progress on the monitor of the device. An RC circuit with a 10 kΩ resistor and a 1 µF capacitor will suffice.

10 More Experiments with the Arduino

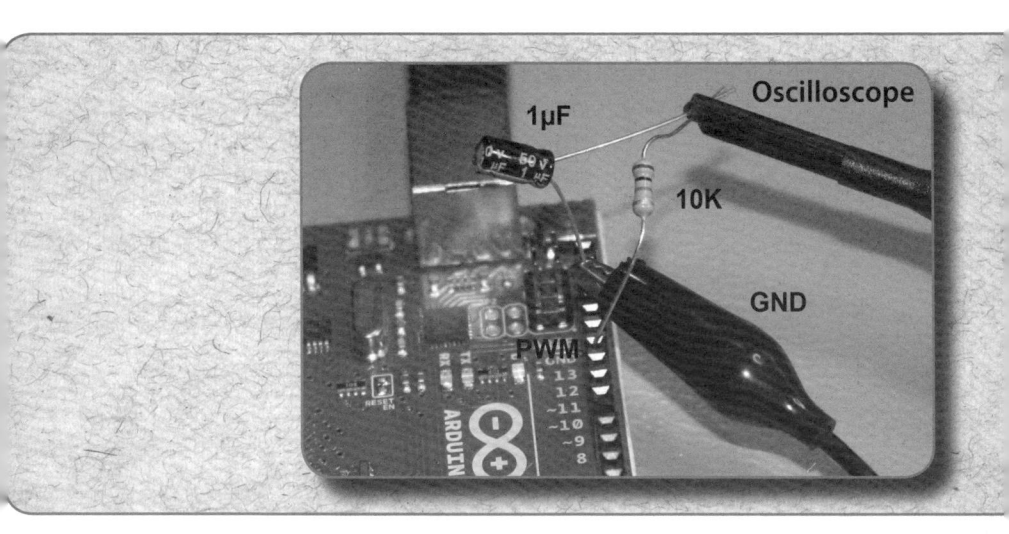

Figure 10.3: Set-up of the RC circuit. The resistor is connected to analog output 11 on the Arduino board. The negative terminal of the capacitor is attached to ground. This circuit filters the PWM signal so that only the envelope of the sine function is visible on the oscilloscope.

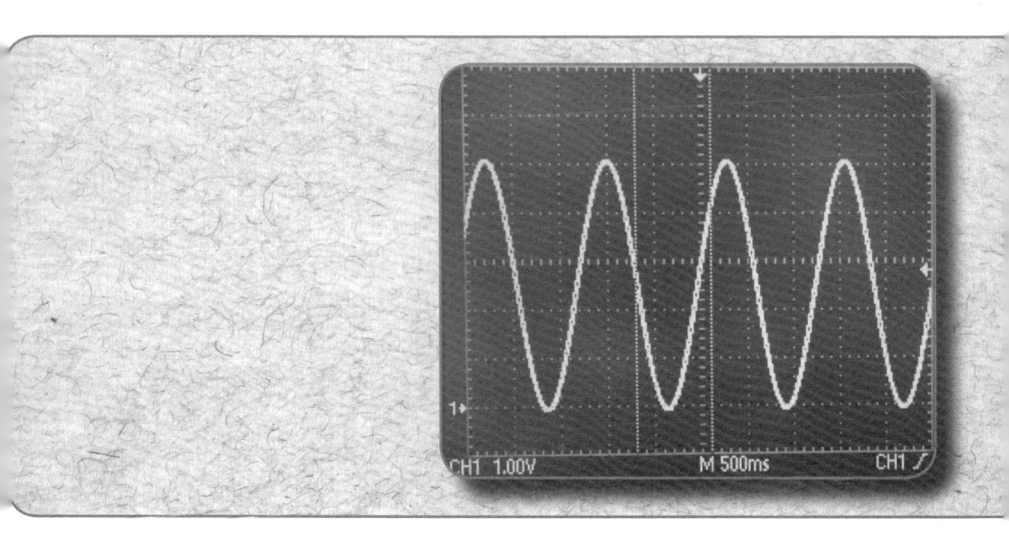

Figure 10.4: The result on the oscilloscope after attaching the RC circuit

Tip More information about envelopes of signals and RC circuits can be found at:

http://en.wikipedia.org/wiki/Envelope_detector
http://en.wikipedia.org/wiki/RC_circuit

Calculating the sine function in the program

The second example shows how you can calculate the sine function for the sine wave blinker in the program. In the previous example, we have used a table, now we will carry out the calculation directly on the microcontroller. The program is much smaller, but the calculation puts a lot of stress on the microcontroller so that the run time is significantly increased.

The `sin()` function accepts a value in radians. First you have to convert angular degrees to radians, which is done by `x*(pi/180)`. When you multiply the result with 255 (PWM range from 0 to 255) you scale the sine function to the range from 0 to 255.

Example: Sine wave blinker with sine function

Time needed: 10 min
Difficulty: 2

```
// Franzis Arduino
// Soft blinker with sine function

int ledPin = 11;
float Val;
int led;

void setup()
{
    pinMode(ledPin, OUTPUT);
}
```

```
void loop()
{
    for (int x=0; x<180; x++)
    {
        Val = (sin(x*(3.1416/180)));
        led = int(Val*255);
        analogWrite(ledPin, led);
        delay(10);
    }
}
```

10.3 | Debouncing buttons

Due to their mechanical composition, push-buttons exhibit the characteristic trait of »bouncing«. Whenever you press or release the button, the signal does not change immediately to *high* or *low*, but issues short impulses that give the impression of someone rapidly operating the button.

How to debounce push-buttons via the software

As these impulses are too short, you cannot see this effect when you use the button to switch on or off a light bulb. However, the controller retrieves the button state so fast that he gains the impression of a button that is rapidly pressed and released. In order to determine a steady state, the button has to be debounced by means of the software.

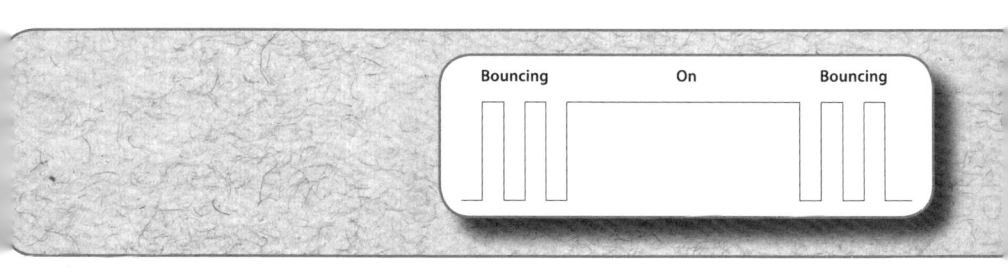

Figure 10.5: This is the impression the microcontroller gains at the digital input due to the bouncing effect of a button

The problem can be avoided by retrieving the state two times with a short delay between the readings. Only when the signal level at the second reading is identical to the level at the first try, you can act on the assumption that the button was actually pressed (or depressed) and the current digital value on the input is correct. The delay should be in the range from 20 to 100 ms.

Figure 10.6: Diagram of the set-up for an LED dimmer with transistor

Required parts for the experiment
- 1 x microcontroller board Arduino Uno
- 1 x breadboard
- 1 x push-buttons
- 2 x jump wire, ca. 5 cm

Example: Debouncing a push-button V1

```
// Franzis Arduino
// Debouncing a push-button V1

int SW1=12;

void setup()
{
   Serial.begin(9600);
   pinMode(SW1,INPUT);
   digitalWrite(SW1,HIGH);
   Serial.println("Debouncing a push-button V1");
}

void loop()
{
   if(!digitalRead(SW1))
   {
      delay(50);
      if(!digitalRead(SW1))
      {
         Serial.println("Button SW1 has been
                        pressed");
      }
   }
}
```

Time needed: 10 min
Difficulty: 2

■ ■ ☐ ☐ ☐

In this code, text is output to Terminal when the button is pressed. The state of the button is read (query for a `low` state), and after a short delay (50 ms) it is read again. If it is still `low`, the text is printed to Terminal.

The drawback of this method is, that the programs is called so often until the button is released. Another possibility is to execute the program code and then wait for

Simple debouncing

the button to be released. The program runs through the `while(!digitalRead(SW1));` loop until the button is pressed no longer.

Example: Debouncing a push-button V2

Time needed: 10 min
Difficulty: 2

```
// Franzis Arduino
// Debouncing a push-button V2

byte i=0;
int SW1=12;

void setup()
{
  Serial.begin(9600);
  pinMode(SW1,INPUT);
  digitalWrite(SW1,HIGH);
  Serial.println("Debouncing a push-button V2");
}

void loop()
{
  if(!digitalRead(SW1))
  {
    delay(50);
    if(!digitalRead(SW1))
    {
      i++;
      Serial.print("Button SW1 was pressed ");
      Serial.print(i,DEC);
      Serial.println("x times");
      do{
      }while(!digitalRead(SW1));
    }
  }
}
```

10 More Experiments with the Arduino

The reverse behaviour can be obtained by placing the do while loop at the beginning. Now the code will be executed only after the button is released.

Example: Debouncing a push-button V3

Time needed: 10 min
Difficulty: 2

```
// Franzis Arduino
// Debouncing a push-button V3

byte i=0;
int SW1=12;

void setup()
{
  Serial.begin(9600);
  pinMode(SW1,INPUT);
  digitalWrite(SW1,HIGH);
  Serial.println("Debouncing a push-button V3");
}

void loop()
{
  if(!digitalRead(SW1))
  {
    delay(50);
    if(!digitalRead(SW1))
    {
      do{
      }while(!digitalRead(SW1));
      i++;
      Serial.print("Button SW1 was pressed ");
      Serial.print(i,DEC);
      Serial.println("x times");
    }
  }
}
```

Debouncing for advanced learners

The following code provides an even better (and nearly perfect) solution. It is an amalgamation of the previous example. Furthermore, the results are not only retrieved twice but also compared. The value of `digitalRead` has to be identical at two points in a given time period in order to run the code. As a further addition, we turn on or off LED L on the Arduino board.

Example: Debouncing a push-button V4

Time needed: 10 min
Difficulty: 3

```
// Franzis Arduino
// Debouncing a push-button V4

byte i=0;
int SW1=12;
int LED=13;
int TOG=0;
byte value_1, value_2=0;

void setup()
{
    Serial.begin(9600);
    pinMode(SW1,INPUT);
    digitalWrite(SW1,HIGH);
    pinMode(LED,OUTPUT);
    Serial.println("Debouncing a push-button V4");
}

void loop()
{
    value_1=digitalRead(SW1);
    if(!value_1)
    {
        delay(50);
        value_2=digitalRead(SW1);
```

```
if(!value_2)
{
    i++;
    Serial.print("Button SW1 was pressed ");
    Serial.print(i,DEC);
    Serial.println("x times");
    if(TOG!=0)TOG=0;else TOG=1;
    digitalWrite(LED,TOG);
    do{
    }while(!digitalRead(SW1));
}
}
```

0.4 | A simple switch-on delay

s the name implies, a switch-on delay switches on a con-
ımer (in our case, the LED L) with a delay after pressing
ıe button. In our example, the delay is implemented by
ıe `delay()` command and a counting loop. When you
ress the button, a flag stores the state und increments
ıe variable `i`. When `i` exceeds the preset amount of mil-
;econds (in this case 3000 ms or 3 s), LED L is turned on
ıd the program gets »trapped« in the `while(1)` loop.

A simple switch-on delay switches on a consumer with a delay after pressing the button

As in the examples about the debouncing of push-but-
ɔns, the button is attached to digital pin 12 and ground. Now
ɔu have to press the push-button once and then release
 in order to leave `do{}while(!digitalRead(SW1));`.

In querying the button state, the flag is set to 1. Now
ıe incrementation of the variable `i` begins. When it ex-
:eds 3000, the LED is turned on. Due to `delay(1)`, the
crease of `i` by 1 only happens every millisecond.

Example: Switch-on delay

Time needed: 10 min
Difficulty: 2
■ ■ ☐ ☐ ☐

```
// Franzis Arduino
// Switch-on delay

int SW1=12;
int value_1, value_2=0;
int LED=13;
byte Flag=0;
int i=0;

void setup()
{
   pinMode(SW1,INPUT);
   digitalWrite(SW1,HIGH);
   pinMode(LED,OUTPUT);
}

void loop()
{
   value_1=digitalRead(SW1);

   if(!value_1)
   {
      delay(50);
      value_2=digitalRead(SW1);
      if(!value_2)
      {
         Flag=1;
         do{
         }while(!digitalRead(SW1));
      }
   }
```

10 More Experiments with the Arduino 197

```
    if(Flag==1)i++;
    if(i>3000)
    {
        digitalWrite(LED,HIGH);
        while(1);
    }
    delay(1);
}
```

0.5 | A simple switch-off delay

The counterpart of the switch-on delay is the switch-off delay. With this, a consumer is turned off with a preset delay after pressing the button. The procedure is identical to that of the switch-on delay, but here the variable i is decremented instead of incremented.

A switch-off delay switches off a consumer after a preset time

Example: Switch-off delay

```
// Franzis Arduino
// Switch-off delay

int SW1=12;
int value_1, value_2=0;
int LED=13;
byte Flag=0;
int i=3000;

void setup()
{
    pinMode(SW1,INPUT);
    digitalWrite(SW1,HIGH);
```

Time needed: 10 min
Difficulty: 2
■ ■ □ □ □

```
   pinMode(LED,OUTPUT);
   digitalWrite(LED,HIGH);
}

void loop()
{
   value_1=digitalRead(SW1);

   if(!value_1)
   {
      delay(50);
      value_2=digitalRead(SW1);
      if(!value_2)
      {
         Flag=1;
         do{
         }while(!digitalRead(SW1));
      }
   }

   if(Flag==1)i--;
   if(i==0)
   {
      digitalWrite(LED,LOW);
      while(1);
   }
   delay(1);
}
```

10 More Experiments with the Arduino

10.6 | LEDs

Calculating the series resistors for LEDs

In most of the previously described applications, one or more LEDs were used as output to test the software. You may have asked yourself how you have to calculate the series resistor in cases like these.

An LED is very much like a normal silicon diode, but it is operated in conducting direction (anode to the positive pole and cathode to the negative). There is a voltage drop along the LED, the amount of which depends on the colour (between 1.6 and 4 V).

The exact voltage is given in the data sheet for the LED and is called $V_{forward}$. The LED also needs some current so that it can light up. This current is called $I_{forward}$ in the data sheets. In this tutorial kit, we only use low-current LEDs with a maximum operating current of 2 mA.

An example for the calculation:

$I_{forward}$ = 2 mA (low-current LED)
$V_{forward}$ = 2.2 V
Operating voltage of the Arduino Vcc = 5 V
R = x Ω (the quantity we want to determine)

$$R = \frac{VCC - V_{forward}}{I_{forward}} = \frac{5V - 2.2V}{2mA} = 1400 \, \Omega$$

Use a series resistor of the E12 series with a little higher value, namely 1.5 kΩ, to make sure the LED will not be damaged.

| Franzis Arduino Tutorial Kit

Double-flash LED signal

As a hands-on example, we will build an LED double flasher: The LEDs attached to the digital pins 10 and 11 blink alternately three times each. This simulates the light effect of the beacon light on an ambulance car.

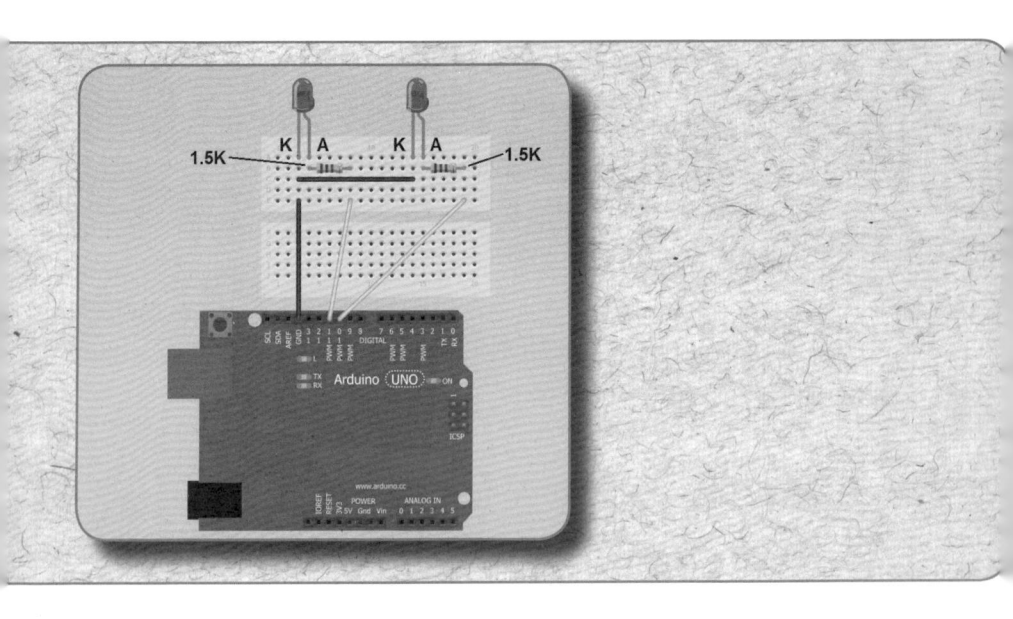

Figure 10.7: Diagram of the circuit

Required parts for the experiment
- 1 x microcontroller board Arduino Uno
- 1 x breadboard
- 2 x red LED
- 2 x 1.5 kΩ resistors
- 4 x jump wire, ca. 10 cm

Example: *Double flasher*

Time needed: 10 min
Difficulty: 3
■ ■ ■ ☐ ☐

```
// Franzis Arduino
// Double flasher

int LED_1=10;
int LED_2=11;
int i=0;
int TOG=0;

void setup()
{
    pinMode(LED_1,OUTPUT);
    pinMode(LED_2,OUTPUT);
}

void loop()
{
    for(i=0;i<3;i++)
    {
        if(TOG==0)TOG=HIGH;else TOG=LOW;
        digitalWrite(LED_1,TOG);
        delay(40);
    }
    TOG=0
    digitalWrite(LED_1,LOW);
    delay(100);

    for(i=0;i<3;i++)
    {
        if(TOG==0)TOG=HIGH;else TOG=LOW;
        digitalWrite(LED_2,TOG);
        delay(40);
    }
    digitalWrite(LED_2,LOW);
    delay(100);
```

```
for(i=0;i<3;i++)
{
    if(TOG==0)TOG=HIGH;else TOG=LOW;
    digitalWrite(LED_1,TOG);
    delay(40);
}

digitalWrite(LED_1,LOW);
delay(500);
}
```

The program runs through the first `for` loop and lets the LED at digital pin 10 blink three times. Then it enters the second `for` loop and causes the second LED to blink three times. After that, it waits for 500 ms and starts again.

10.7 | Switching large consumers

Operating mode of a transistor

If you need more current than our port can provide (max ±40 mA), you will have to amplify it by a transistor as you did in the dimmer project. Let us have a more detailed look at transistors and their properties.

In a transistor, a small current (I_B) flows to the base and provides for a larger collector current (I_C). The amplification (expressed as the so called h_{FE} value) of small-signal transistors amounts to a factor of 100 to 1000, depending on the model. The transistor BC548C that we use in our experiments has an average amplification factor of about 300. A base current of 0.1 mA will therefore result in a collector current of 30 mA. The collector current for our transistor must not exceed 100 mA. Again, we will use an LED with a series resistor for demonstration purposes.

10 More Experiments with the Arduino 203

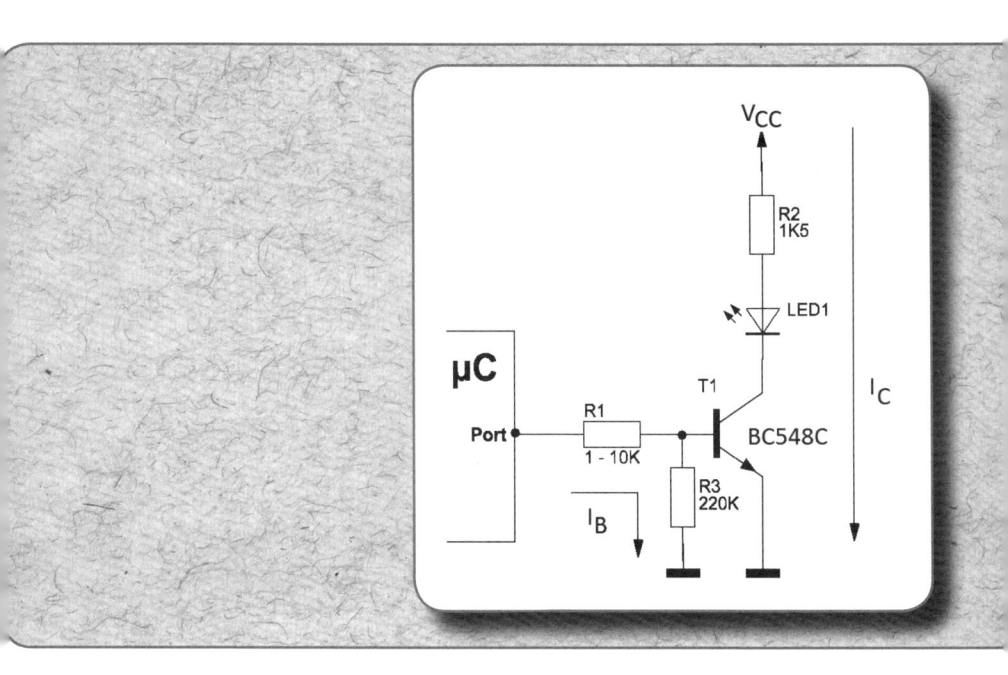

Figure 10.8: A transistor at the digital output of the Arduino microcontroller (µC); I_B = base current, I_C = collector current. Both base and collector current flow through the emitter. The tutorial kit contains low-current LEDs ($I_{forward}$ = 2 mA), hence the large 1.5 kΩ series resistor R2.

For the R1 resistor, we choose a value between 1 and 10 kΩ, depending on the application. With a BC548C transistor a 10 kΩ resistor is sufficient to fully illuminate an LED.

The resistor R3 serves to protect the base against interference. When you switch on the Arduino, the digital pins have a high resistance because they are initialized as inputs. The base would be »up in the air«. To avoid this, we attach a 220 to 470 kΩ resistor directly to the base and against ground. This makes sure that the transistor connects through only when a larger current flows to the base.

The resistor serves to protect the base against interference

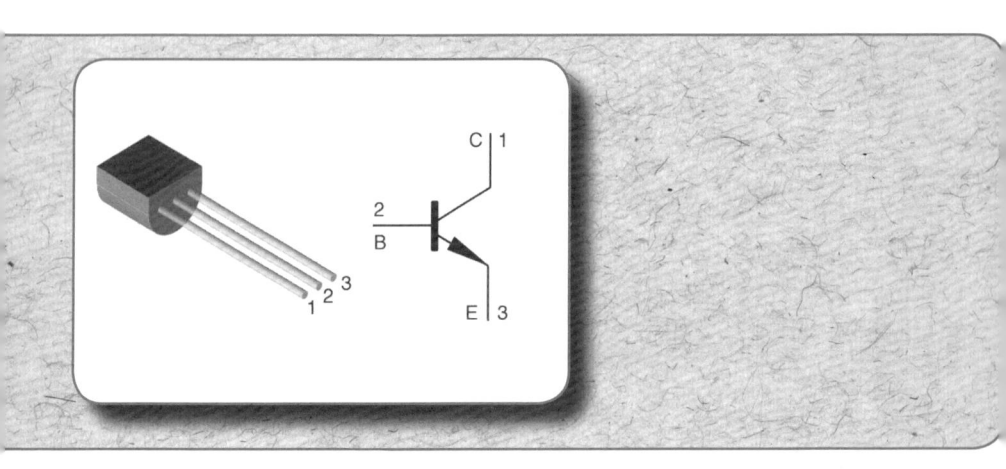

Figure 10.9: Pin configuration of the BC548 transistor (Source: Vishay data sheet)

The more current the consumer needs, the more current must flow to the base so that a larger collector current is possible.

How to calculate the collector current

The collector current is calculated as follows:

$$I_C = I_B \times h_{FE} \text{ (transistor amplification factor)}$$

The following circuit diagram shows a transistor controlling a small relay. The resistance value of resistor R5 may be 1 to 22 kΩ, depending on the coil current. In general you can use a 4.7 kΩ resistor because the transistor works as a simple switch.

There is no interference-suppression resistor in this circuit, but you can add one in case you experience any problems. As in the previous example, you can use a 220 kΩ resistor between base and ground.

10 More Experiments with the Arduino

Button S1 uses R4 as an external pull-up resistor, where R4 should have a resistance between 10 and 22 kΩ. Diode D1 prevents the inductive voltage in the relay coil from damaging the transistor when switching off. The inductive voltage is polarised in the opposite direction of the source. Thus the diode has to be inserted in a way that short-circuits the inductive voltage. In this example, the relay turns on lamp La1 when the digital pin is *high* (5 V).

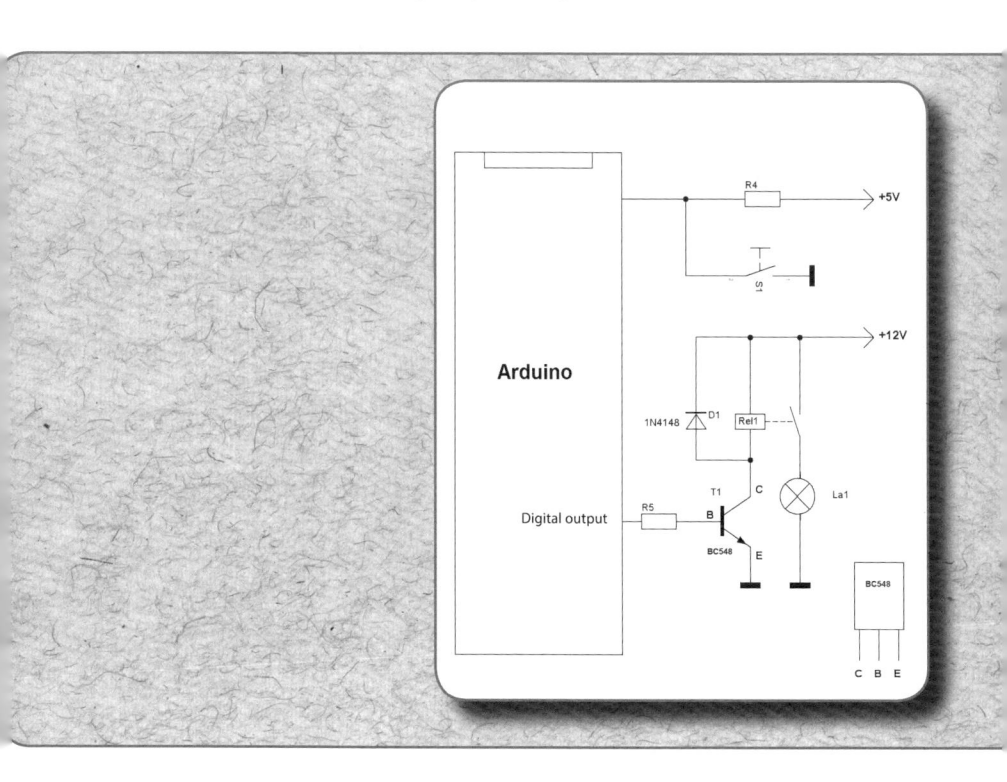

Figure 10.10: Relay at the digital pin of the Arduino

There are many different models of relays. They all have potential-free contacts, i.e., the contact has no connection to the microcontroller circuit whatsoever.

Relays have a potential-free contact

Figure 10.11: Low-power print relay (Source: Conrad Electronic SE)

10.8 | Using the PWM Pins as DAC

Digital/analog conversion

Most naturally occuring signals are analog. For a digital machine that has to control external processes it is therefore very important to translate digital values to analog quantities. When you need an analog voltage, you have to attach an RC circuit to the analog output. This circuit transforms the PWM signal to a quasi-analog voltage.

Most microcontrollers do not have an integrated DAC

Most microcontrollers, including the one on the Arduino, do not have an integrated digital/analog converter (DAC). It is, however, possible to implement a DAC by means of PWM signals. This way you can generate an adjustable direct-current voltage.

10 More Experiments with the Arduino

Running a PWM signal through a low-pass filter results in a direct-current voltage with an alternating component. The mean value is proportional to the duty cycle of the PWM signal.

Running a PWM signal through a low-pass filter results in a direct-current voltage with an alternating component

The important thing for the task at hand is the fact that a duty cycle of 50% nearly exactly results in half of the original PWM output voltage. When we increase the duty cycle, the output voltage behind the low-pass filters increases, too, and vice versa.

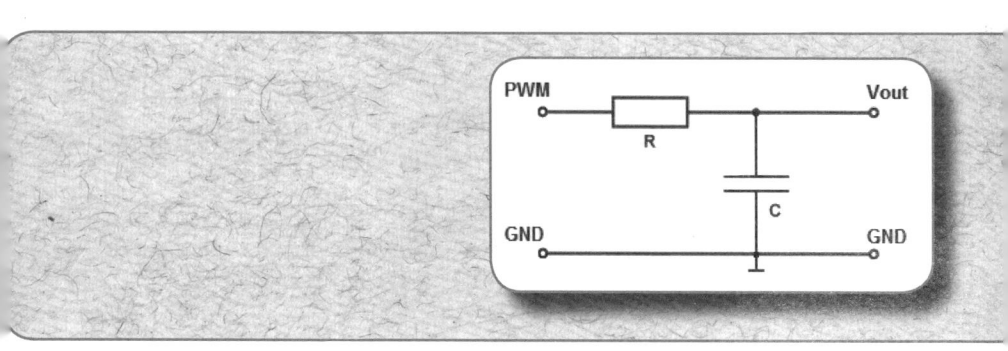

Figure 10.12: Set-up of an RC low-pass filter to generate an analog voltage out of a PWM output

The quantities of the RC circuit can be calculated as follows:

$$f_G = \frac{1}{2 \times \pi \times R \times C}$$

How to calculate the RC circuit

f_G = cut-off frequency of the PWM signal
π = 3.1416
R = resistance in Ohm
C = capacitance in Farad

The RC low-pass filter used in the example consists of a 1 µF capacitor and a 10 kΩ resistor and has a cut-off frequency of about 15 Hz.

This simple RC low-pass filter must not be subjected to high loads, because otherwise the capacitor would discharge too quickly. This would render the filter useless and increase the ripple. It is best to put a current output stage behind the RC circuit in order to raise the signal to the desired voltage level and to detach the low-pass filter from the rest of the circuit.

Tip More information on PWM and DAC can be found at

http://www.mikrocontroller.net/articles/Pulsweitenmodulation

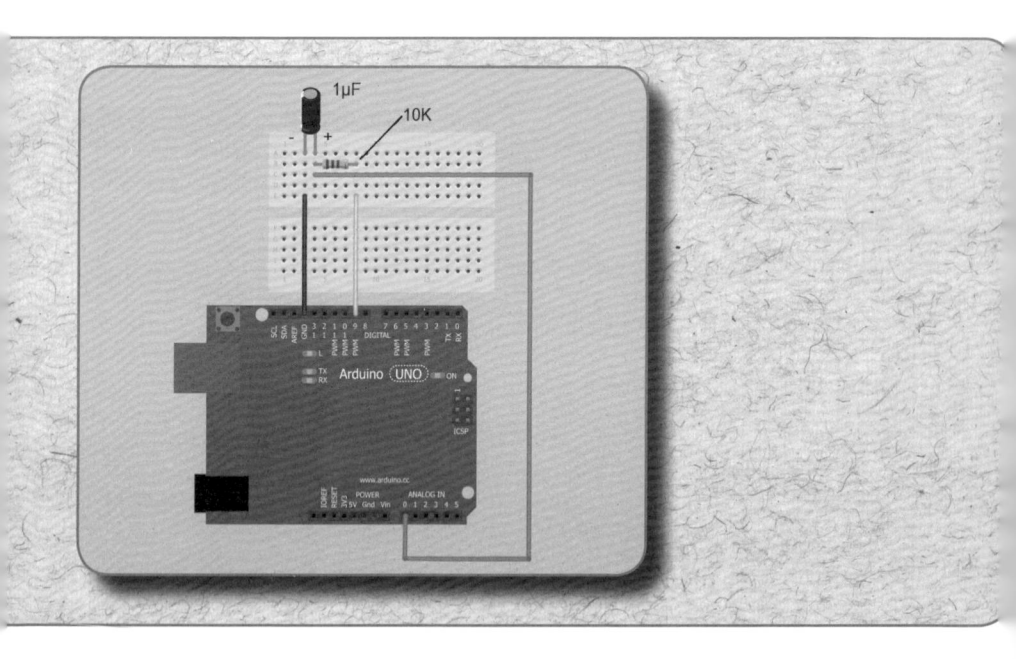

Figure 10.13: Diagram of the circuit used to generate an analog voltage via an RC low-pass filter and to read it at the analog input 0

10 More Experiments with the Arduino

Caution Pay attention to the polarity of the capacitor. The white bar marks the ground pin.

Pay attention to the polarity of the capacitor!

- 1 x microcontroller board Arduino Uno
- 1 x breadboard
- 1 x 1 µF capacitor
- 1 x 10 kΩ resistor
- 2 x jump wire, ca. 5 cm
- 1 x jump wire, ca. 10 cm

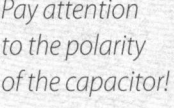

Required parts for the experiment

Example: DAC

```
// Franzis Arduino
// DAC


char buff er[18];
int pinPWM=9;
int raw=0;
float Volt=0;


void setup()
{
    Serial.begin(9600);
    Serial.println("PWM output as DAC");
    Serial.println();
    Serial.println("Enter a value between 0
                    and 255");
    Serial.flush();
}


void loop()
{
    if (Serial.available() > 0)
    {
        int index=0;
```

Time needed: 15 min
Difficulty: 3

■ ■ ■ ☐ ☐

```
        delay(100);
        // Waits until the characters are in the buffer
        int numChar = Serial.available();
        if (numChar>15)
        {
           numChar=15;
        }
        while (numChar--)
        {
           buffer[index++] = Serial.read();
        }
        splitString(buffer);
    }
}

void splitString(char* data)
{
   Serial.print("The following value is received: ");
   Serial.println(data);
   char* parameter;
   parameter = strtok (data, " ,");
   while (parameter != NULL)
   {
      setPWM(parameter);
      parameter = strtok (NULL, " ,");
   }
   // Clears the puffer
   for (int x=0; x<16; x++)
   {
      buffer[x]='\0';
   }
   Serial.flush();
}
```

```
void setPWM(char* data)
{
    int Ans = strtol(data, NULL, 10);
    Ans = constrain(Ans,0,255);
    analogWrite(pinPWM, Ans);
    Serial.print("PWM = ");
    Serial.println(Ans);
    delay(50);
    raw=analogRead(0);
    float ref=5.0/1024.0;
    Volt=raw*ref;
    Serial.print("The voltage at ADC0 is: ");
    Serial.print(Volt);
    Serial.println(" Volt");
    Serial.println();
    Serial.println("Enter a value between 0 and 255");
}
```

This program generates a PWM signal at the analog output pin 9. This signal is then smoothed in the RC low-pass filter and read at the analog input 0. The smaller the value entered by the user, the smaller the output voltage at the filter. The time for a change from 0 to 5 V or vice versa is ca. 40 ms.

This example program also shows how to provide for data input via the serial interface and how to use some of the statements you have learned so far. Study the code thoroughly and look up the statements in the chapter on the programming basics, if necessary.

Data input via the serial interface

A new aspect is the data type `char*` in the `void splitString(char* data)` function. When you pass an array to a function, you have to specify the data type marked by an asterisk (here `char*`) and then the name (here `data`) of the array in question.

The functions `strtok` and `strtol` are C++ string operations for delimiting characters. These routines can do much more than we need in this example.

In the routines for receiving serial data, you can also transfer and process several values at the same time, but you have to delimit them by a comma. However, this is not necessary in our example. More information on C++ functions, some of which can be used in Arduino programming, can be found in the reference at:

http://www.cplusplus.com/reference/cstring/strtok/

and:

http://playground.arduino.cc/Interfacing/CPPWindows

10.9 | Music's in the air

Using a piezo buzzer with the Arduino

Among other things, the Arduino is a »musical« device. To determine how far this »musicality« goes, we attach the little piezo buzzer in the tutorial kit to the Arduino board, as shown in the figure.

Attach the piezo buzzer to digital output 11 and ground.

Use tone to generate sounds

For generating sounds at a pin, the Arduino IDE provides the `tone()` command. It toggles the *high/low* state of the specified pin at a specified frequency. You have seen this before in the PWM examples, but here the frequency is generated solely by the software and not by means of the microcontroller hardware. As long as the sound is issued, the program pauses.

10 More Experiments with the Arduino

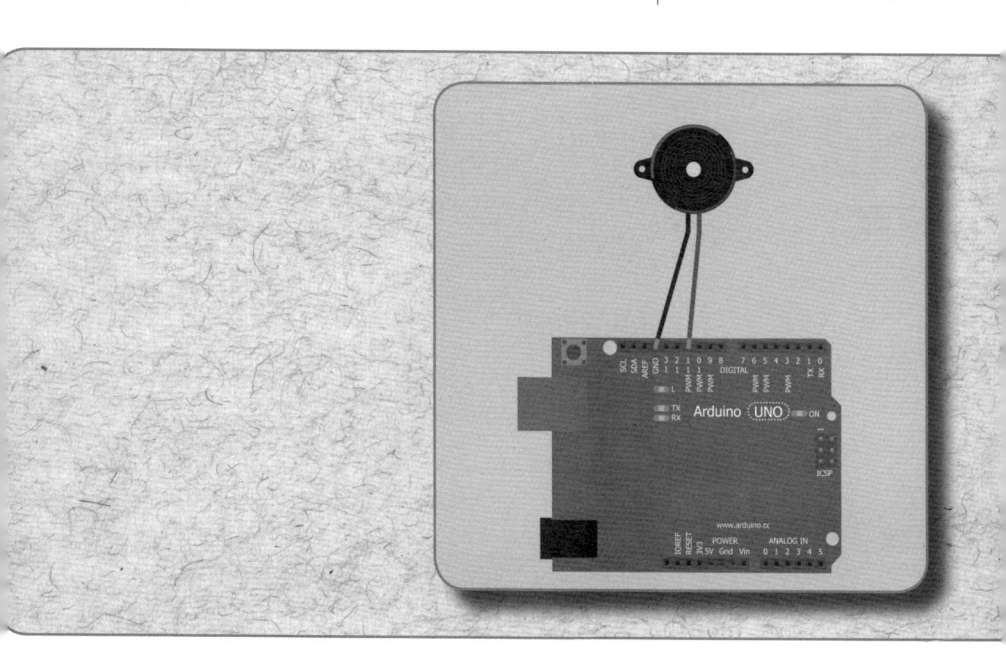

Figure 10.14: Set-up with piezo buzzer

- 11 x microcontroller board Arduino Uno
- 1 x piezo buzzer

Required parts for the experiment

The `tone()` statements needs several parameters: the pin to which the buzzer is attached, the frequency specified in Hertz (i.e. the number of pulses per second), and the duration of the sound.

```
tone(pin, frequency);
    // Issues a continuous sound
tone(pin, frequency, duration);
    // Issues a sound for the specified time
```

Time needed: 5 min
Difficulty: 2
■ ■ ☐ ☐ ☐

Example: Sound

```
// Franzis Arduino
// Sound

int Speaker=11;

void setup()
{
    pinMode(Speaker, OUTPUT);
}

void loop()
{
    tone(Speaker,550,450);
    delay(3000);
}
```

Use tone for switching sounds

The `tone()` command is only intended for switching tones and very simple melodies. It is not possible to generate an exact frequency.

The following example shows how to write a custom sound routine emulating the functionality of `tone()`. Again, the sounds are generated by rapidly toggling the state of a digital pin from `high` to `low` and back.

```
digitalWrite(Speaker, HIGH);
delayMicroseconds(tone);
digitalWrite(Speaker, LOW);
delayMicroseconds(tone);
```

When you run this code inside an infinite loop, it rapidly toggles the state of the `Speaker` pin from `high` to `low` and back and thus generates a frequency at the digital output. The period is determined by the variable `tone`

10 More Experiments with the Arduino

that defines the delay in microseconds after setting the pin from `high` to `low` and before setting it back to `high`.

```
Frequency in Hertz (Hz) = 1/period in seconds
```

Example: Melody

Time needed: 10 min
Difficulty: 3

```
// Franzis Arduino
// Melody

int Speaker = 11;
int length = 15;
char notes[] = "ccggaagffeeddc ";
int beats[] = { 1, 1, 1, 1, 1, 1, 2, 1, 1, 1, 1,
1, 1, 2, 4 };
int tempo = 300;

void setup()
{
    pinMode(Speaker, OUTPUT);
}

void loop()
{
    for (int i = 0; i < length; i++)
    {
        if (notes[i] == ' ')
        {
            delay(beats[i] * tempo);
        }
        else
        {
            playNote(notes[i], beats[i] * tempo);
        }
        delay(tempo / 2);
    }
}
```

```
void playTone(int tone, int duration)
{
    for (long i = 0; i < duration * 1000L; i +=
tone * 2)
    {
        digitalWrite(Speaker, HIGH);
        delayMicroseconds(tone);
        digitalWrite(Speaker, LOW);
        delayMicroseconds(tone);
    }
}

void playNote(char note, int duration)
{
    char names[] = { 'c', 'd', 'e', 'f', 'g', 'a',
'b', 'C' };
    int tones[] = { 1915, 1700, 1519, 1432, 1275,
1136, 1014, 956 };
    for (int i = 0; i < 8; i++)
    {
        if (names[i] == note)
        {
            playTone(tones[i], duration);
        }
    }
}
```

This a relatively complex example that uses many arrays. A closer look shows that these are mostly value assignments via the index carried out in loops. For instance every note (c, d, e, etc.) is assigned a frequency (1015, 1700, 1519, etc.) in the playNote() function. In the arrays, the values are organized in such a way that an assignment via the index is possible. Thus note C gets the value 1915 via index 0.

Further down in the code, there is something new: the suffix L after `1000` (`1000L`). It forces the compiler to use `long` variables for the calculations in order to avoid any overflows.

Tip The sound gets considerably louder when you attach a party blowout to the piezo buzzer. Another possibility is to put the buzzer onto a piece of paper that acts as a speaker membrane.

How to amplify the sounds

10.10 | Romantic Candlelight, Courtesy of the Microcontroller

We all love the cozy evenings illuminated by candles that flicker in the wind and evoke a romantic atmosphere. You can recreate this atmosphere and simulate flickering candlelight with an Arduino, three LEDs, and a random number generator.

The Arduino goes romantic

Let us see how to make candlelight by means of the Arduino. This sort of light is especially beautiful when used for a model railroad or a nativity scene. For miniature worlds like these, you always need some light effects to simulate the real thing.

The program uses the analog outputs of the microcontroller. This time, we do not preset a steady value or a sine wave, but determine the duty cycle via `random()`, so that the brightness of the red and the yellow LEDs will always vary. When you put the LEDs into a jar made of frosted glass and cover the inside of the lid with aluminum foil, the light will be dispersed equally. This way, you

Light effects for floor lamps

can build a beautiful eye-catcher for your patio or your living room. When you put it in the right place, the differences between real candles and the LED simulation will vanish.

It is even simpler to build a small box of paper and put it over the LEDs. The edges of the box should each be ca 10 cm long. You can feed the Arduino by a battery and put it into the box as well. When you attach transistors as amplifiers to the outputs and use high-power LEDs, you can even integrate the set-up into a floor lamp for a beautiful illumination.

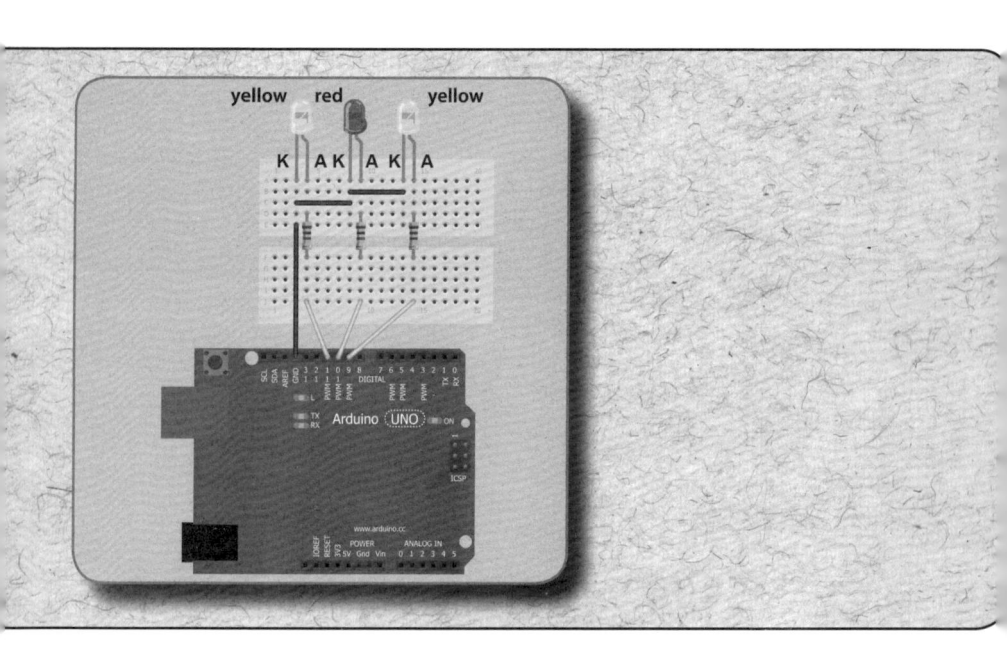

Figure 10.15: Set-up of the microcontroller candlelight

10 More Experiments with the Arduino

- 1 x microcontroller board Arduino Uno
- 1 x breadboard
- 1 x red LED
- 1 x yellow LED
- 3 x 1.5 kΩ resistors
- 3 x jump wire, ca. 5 cm
- 1 x jump wire, ca. 10 cm

Required parts for the experiment

Example: Candlelight

Time needed: 10 min
Difficulty: 3

```
// Franzis Arduino
// Candlelight

int led_yellow1 = 9;
int led_red = 10;
int led_yellow2 = 11;

void setup()
{
    pinMode(led_yellow1, OUTPUT);
    pinMode(led_red, OUTPUT);
    pinMode(led_yellow2, OUTPUT);
}

void loop()
{
    analogWrite(led_yellow1, random(120)+135);
    analogWrite(led_red, random(120)+135);
    analogWrite(led_yellow2, random(120)+135);
    delay(random(100));
}
```

10.11 | Surveillance at the Exit for Staff Members

Perfect theft protection using a random number generator

In many companies it is necessary to check staff members for stolen goods. Normally this is done by the security staff. When you do not want the guards to select the staff members to be searched, you can use a random number generator to make the decision. A system that randomly filters out single employees for briefcase inspection has a certain deterrent effect.

The concept is relatively simple: A push-button is installed at the exit. When leaving the premises, every staff member has to push that button. A random number generator determines when to turn on a red light and a buzzer to send for a security guard that will inspect the briefcase. After that, the random number generator calculates a new number, and the whole procedure starts again. The sensitivity can be adjusted by a potentiometer. The greater the value at analog input 0, the less alarms are triggered, because the range and the related variance increase.

Required parts for the experiment
- 1 x microcontroller board Arduino Uno
- 1 x breadboard
- 1 x red LED
- 1 x piezo buzzer
- 1 x 1.5 kΩ resistors
- 1 x push-button
- 4 x jump wire, ca. 5 cm

10 More Experiments with the Arduino | 221

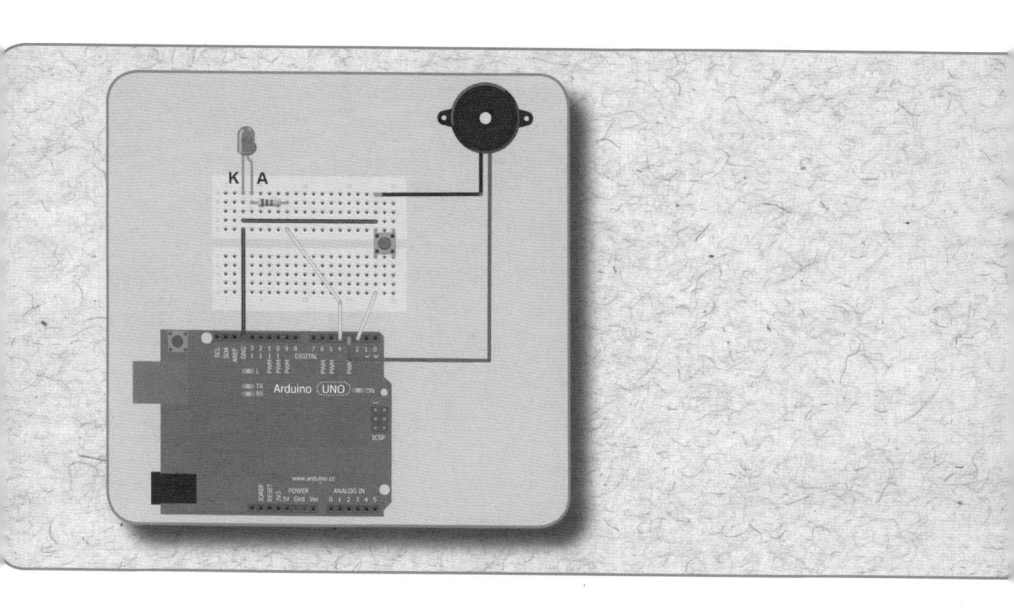

Figure 10.16: Set-up of the circuit

Example: Exit for staff members

```
// Franzis Arduino
// Surveillance at the exit for staff members

int i , x=0;
int LED=4;
int SW1=2;
int Sensitivity=0;
int Speaker=3;
int Person=0;

void setup()
{
    pinMode(LED,OUTPUT);
    pinMode(Speaker,OUTPUT);
```

Time needed: 20 min
Difficulty: 2

```
    pinMode(SW1,INPUT);
    digitalWrite(SW1,HIGH);
    randomSeed(1000);
}

void loop()
{
    Person=(77+analogRead(Sensitivity)/10);
    i=random(1,Person);
    while(1)
    {
        if(!digitalRead(SW1))
        {
            delay(50);
            if(!digitalRead(SW1))
            {
                if(x>Person)x=0;
                if(i==x)
                {
                    digitalWrite(LED,HIGH);
                    tone(Speaker,500,250);
                    delay(3000);
                    digitalWrite(LED,LOW);
                    break;
                }
                x++;
            }
        }
    }
}
```

0.12 | An Arduino Clock

Many applications need a clock to control the program. This can be a simple timer, a scheduling control or an hour meter. The output to Terminal occurs every second. When the counter reaches 100 (`cnt = 100`), the time is printed. As a means of monitoring the program we let LED L blink every second.

Keep in mind that this clock has not the precision of real quartz watch. The microcontroller has a clock frequency of 16 MHz which is much higher than that of a watch (32.768 kHz) and therefore results in a much higher variance. Deviations of more than a minute per day are not uncommon. Furthermore, the precision of the clock depends very much on the ambient temperature. If the temperature varies to much, the clock will exhibit substantial errors. Nevertheless, the clock is suitable for many useful applications. For instance, you can use it as a clock for controlling an irrigation system where to-the-minute precision is not essential.

Time measurement with the Arduino

Tip More information on clock quartzes can be found at:

http://en.wikipedia.org/wiki/Quartz_clock

Example: *Clock*

```
// Franzis Arduino
// Clock

int cnt, Second, Minute, Hour=0;
int LED=13;
```

Time needed: 10 min
Difficulty: 3

■ ■ ■ □ □

```
void setup()
{
  Serial.begin(9600);
  pinMode(LED,OUTPUT);
  // Preset time
  Second=0;
  Minute=0;
  Hour=0;
}

void loop()
{
  cnt++;
  if(cnt==50)digitalWrite(LED,LOW);
  if(cnt==100)
  {
    digitalWrite(LED,HIGH);
    Serial.print(Hour);
    Serial.print(":");
    Serial.print(Minute);
    Serial.print(":");
    Serial.println(Second);
    Second++;
    if(Second==60)
    {
      Second=0;
      Minute++;
      if(Minute==60)
      {
        Minute=0;
        Hour++;
        if(Hour==24)
        {
          Hour=0;
        }
      }
    }
```

```
        cnt=0;
    }
    delay(10);
}
```

10.13 | School Bell Program

A good practical example for a clock is a school bell that rings at preset times (every 45 minutes). To this end, you have to extend the clock program by some line of code to retrieve the time using `if` conditions. Whenever one of the condition holds true, the piezo buzzer issues a sound for five seconds. One possible usage of this program is limiting the time your children are allowed to play on the computer.

Yet another possibility: Using the Arduino as a school bell

In our example however, we pretend that we will really use the bell at school.

At a school, the following schedule may prevail:

1. lesson: 7:00–7:45 hrs
2. lesson: 7:55–8:40 hrs

Morning break: 20 min

3. lesson: 9:00–9:45 hrs
4. lesson: 9:55–10:40 hrs
5. lesson: 10:50–11:35 hrs

Lunch break: 30 min

6. lesson: 12:05–12:50 hrs
7. lesson: 13:00–13:45 hrs (1:00–1:45 p.m.)

At the beginning and the end of each lesson th
school bell shell ring. In our example, the sound is simu
lated by a piezo buzzer.

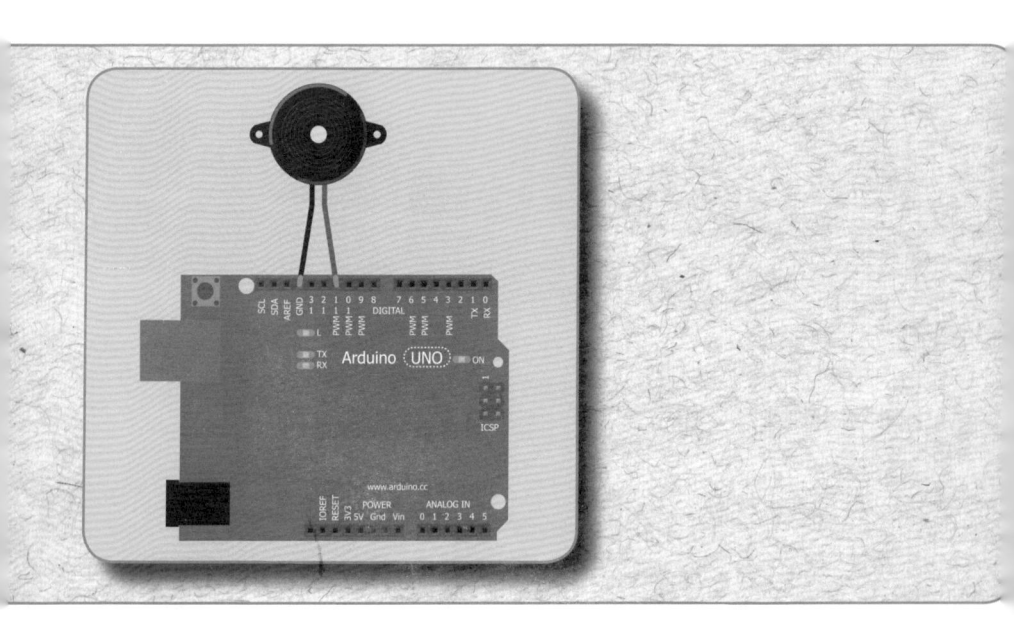

Figure 10.17: Set-up of the school bell experiment; it is identical to the set-up of the melody experiment

Required parts for the experiment
- 1 x microcontroller board Arduino Uno
- 1 x piezo buzzer

Example: School bell

Time needed: 10 min
Difficulty: 4

```
// Franzis Arduino
// School bell

int cnt, Second, Minute, Hour=0;
int LED=13;
int Speaker=11;

void setup()
{
    Serial.begin(9600);
    pinMode(LED,OUTPUT);
    pinMode(Speaker,OUTPUT);
    // Default time
    Hour=6;
    Minute=59;
    Second=58;
}

void loop()
{
    cnt++;
    if(cnt==50)digitalWrite(LED,LOW);
    if(cnt==100)
    {
        digitalWrite(LED,HIGH);
        Serial.print(Hour);
        Serial.print(":");
        Serial.print(Minute);
        Serial.print(":");
        Serial.println(Second);
        Second++;
```

```
    if(Second==60)
    {
        Second=0;
        Minute++;
        if(Minute==60)
        {
            Minute=0;
            Hour++;
            if(Hour==24)
            {
                Hour=0;
            }
        }
    }
    cnt=0;
}
delay(10);
// Ringing times
// 1. lesson
if(Hour==7&&Minute==0)Bell();
if(Hour==7&&Minute==45)Bell();
// 2. lesson
if(Hour==7&&Minute==55)Bell();
if(Hour==8&&Minute==40)Bell();
// Morning break
// 3. lesson
if(Hour==9&&Minute==0)Bell();
if(Hour==9&&Minute==45)Bell();
// 4. lesson
if(Hour==9&&Minute==55)Bell();
if(Hour==10&&Minute==40)Bell();
// 5. lesson
if(Hour==10&&Minute==50)Bell();
if(Hour==11&&Minute==35)Bell();
// Lunch break
```

10 More Experiments with the Arduino

```
// 6. lesson
if(Hour==12&&Minute==05)Bell();
if(Hour==12&&Minute==50)Bell();
// 7. lesson
if(Hour==13&&Minute==0)Bell();
if(Hour==13&&Minute==45)Bell();
}

void Bell(void)
{
  if(Second<5)
  {
    tone(Speaker,500);
  }
  else
  {
    noTone(Speaker);
  }
}
```

Adjusting the clock

At the beginning of the program, the default time for the clock is adopted. As of this moment, the clock starts running. Now the program runs by itself. LED L blinks every second, and the time is transferred to the Terminal on the PC. At the preset times for a lesson, the »bell« is operated for five seconds. A clock that runs too fast or too slow can be adjusted by the `delay()` value. For even more precise adjustment you can use `micros()`. In that case you should monitor the output of LED L on an oscilloscope.

You can configure a separate pin where you tap the 10 ms signal and assign it as output for the oscilloscope or a frequency counter. This way, you can adjust the timing more precisely. It is also possible to measure the run time of the clock with the stop-watch functions that you have already encountered, because the statements in the program need additional time.

However, the stop-watch measurement is derived from the microcontroller quartz that only allows for an estimation of the run time. The clock program even has another blemish: The time is sent to Terminal in the form of »0:0:0« instead of »00:00:00«. As a challenge, try to improve the program!

10.14 | Keypad Lock

Building a keypad lock in no time

You would be no real electronics or programming geek if you were not tempted to secure rooms not accessible to everybody with your own microcontroller keypad lock. As you are now a seasoned Arduino programmer, you can simply build a keypad lock with the components in the tutorial kit!

The lock we present here has only two push-buttons. SW1 at digital input 2 and SW2 at digital input 3. To enter the entrance code, you may have to do something like pressing SW1 twice and SW2 three times. The operation of the buttons is acknowledged by the (red) LED at digital output 4 and the piezo buzzer at analog output 8.

When you have entered the right code, the (green) LED at digital pin 5 is turned on for five seconds. In case of any mistyping you can erase the input by pressing SW2 for a longer time. This is acknowledged by the blinking red LED at digital output 4 and a beeping noise.

10 More Experiments with the Arduino   231

Instead of the green LED, you can attach a door opener via a transistor and a relay, so that you can actually open the door by entering the right code. In our example, the preset code means that you have to press SW1 five times and SW2 three times. Only then the green LED lights up.

The keypad lock also works as a door opener

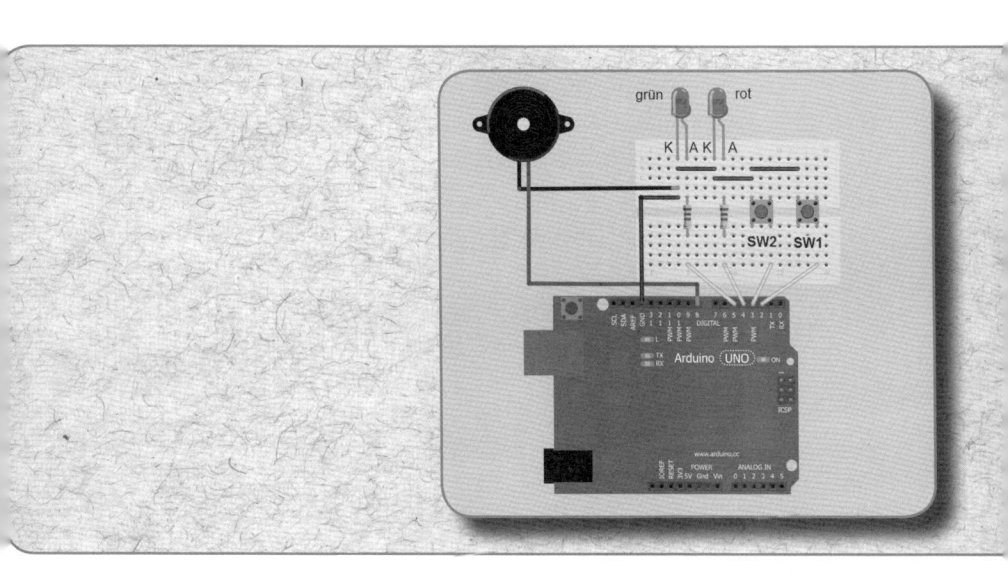

Figure 10.18: Set-up of the keypad lock

- 1 x microcontroller board Arduino Uno
- 1 x breadboard
- 2 x push-buttons
- 2 x LEDs (red and green)
- 1 x piezo buzzer
- 2 x 1.5 kΩ resistor
- 7 x jump wire, ca. 5 cm
- 1 x jump wire, ca. 10 cm

Required parts for the experiment

Time needed: 15 min
Difficulty: 4

Example: Keypad lock

```
// Franzis Arduino
// Keypad lock

int LED_red=4;
int LED_green=5;
int SW1=2;
int SW2=3;
int Buzzer=8;
int x,y,code1,code2,resetTimer=0;

void setup()
{
   pinMode(LED_red,OUTPUT);
   pinMode(LED_green,OUTPUT);
   pinMode(Buzzer,OUTPUT);
   pinMode(SW1,INPUT);
   digitalWrite(SW1,HIGH);
   pinMode(SW2,INPUT);
   digitalWrite(SW2,HIGH);
   Clr_Code();
}

void loop()
{
   // Code 1 = 5
   if(!digitalRead(SW1))
   {
      delay(50);
      if(!digitalRead(SW1))
      {
         Blink();
         x++;
```

```
        if(x==5)
        {
           code1=true;
        }else code1=false;
        do{
        }while(!digitalRead(SW1));
    }
}

// Code 2 = 3
if(!digitalRead(SW2))
{
   delay(50);
   if(!digitalRead(SW2))
   {
      Blink();
      y++;
      if(y==3)
      {
         code2=true;
      }else code2=false;
      do
      {
         delay(50);
         resetTimer++;
         if(resetTimer>50)
         {
            Toggle_Flash();
            Clr_Code();
            break;
         }
      }while(!digitalRead(SW2));
      resetTimer=0;
   }
}
```

```
    if(code1==true&&code2==true)
    {
        digitalWrite(LED_green,HIGH);
        Clr_Code();
        delay(5000);
        digitalWrite(LED_green,LOW);
    }
    else
    {
        digitalWrite(LED_green,LOW);
    }
}

void Blink(void)
{
    digitalWrite(LED_red,HIGH);
    tone(Buzzer,500,150);
    delay(200);
    digitalWrite(LED_red,LOW);
}

void Toggle_Flash(void)
{
    int tog=0;
    for(x=0;x<6;x++)
    {
        if(tog==0)tog=1;else tog=0;
        digitalWrite(LED_red,tog);
        tone(Buzzer,500,250);
        delay(300);
    }
}
```

```
void Clr_Code(void)
{
    x=0;
    y=0;
    code1=0;
    code2=0;
    resetTimer=0;
    delay(1000);
}
```

0.15 | Capacitance meter with auto-range function

Building a capacitance meter for small capacitors

is always an exciting challenge to build measuring devices on your own. With our experimentation board and the programming language Arduino C, we can construct with little effort a capacitance meter for small capacitors in the range from 1 nF to 100 µF.

A capacitance meter with auto-range function works as follows: First the `c_time` variable is set to 0. Digital pin 2 is configured as output and set to `low` to discharge the attached capacitor (the test specimen) before the actual measurement.

After a short discharge delay of one second digital pin 12 is configured as input, and the internal pull-up resistor is activated. This resistor now loads the capacitor until digital pin 12 reads `high`. The time that passed until the `high` level is determined, is measured with the `c_time` variable in the `do-while` loop. `c_time` is proportional to the capacitance of the capacitor. The greater `c_time` is the greater is the capacitance.

To get the actual capacitance value, you have to multiply the variable with a factor. As the detection of a `high` level varies with the microcontroller in question, this factor has to be determined experimentally by measuring several »calibration capacitors«. In a last step, the measurement value is automatically converted to the appropriate range (nanofarad or microfarad) and sent to Terminal. After that, a new measurement can commence.

Formula for converting the measurement values

In order to calibrate the capacitance meter, you have to buy some new capacitors. You can also use the capacitor of the tutorial kit with a known capacitance of 1 µF. necessary, you can ask a technician in an electronics lab to measure your calibration capacitors. However, on most new capacitors, the capacitance value is imprinted (with a tolerance of ±20%, depending on type). For the calibration measurements, you have to use the conversion factor 0 (`c_time*0`).

Insert one of the calibration capacitors in the measuring device and read the value in Terminal. Divide the specified capacitance of the capacitor by the measured value in Terminal and then use the result as the conversion factor. This calculation may look as follows:

1 µF / 19.55 µF = 0.0511

10 More Experiments with the Arduino   237

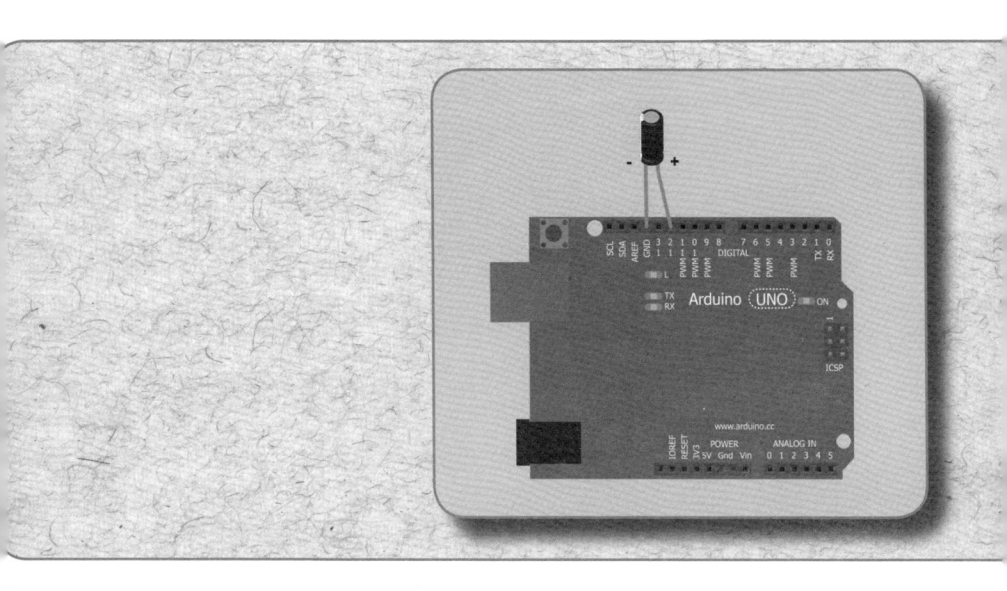

Figure 10.19: Set-up of the capacitance meter

- 1 x microcontroller board Arduino Uno
- 1 x capacitor for testing between 1 nF and 100 µF, min. 5 V

Required parts for the experiment

Caution Always make sure the test capacitor is discharged before you begin with the measurement. The energy of a charged capacitor can destroy the microcontroller!

Also pay attention to the polarization of the capacitor!

Important for the health of your microcontroller!

Example: Capacitance meter

Time needed: 10 min
Difficulty: 3

```
// Franzis Arduino
// Auto-range capacitance meter 1nF to 100 µF

int messPort=12;
float c_time=0.0;
float capacitance=0.0;

void setup()
{
  Serial.begin(9600);
  Serial.println("Auto-range capacitance meter
                  1nF ... 100 uF");
  Serial.println();
}

void loop()
{
  pinMode(messPort,OUTPUT);
  digitalWrite(messPort,LOW);
  c_time=0.0;
  delay(1000);

  pinMode(messPort,INPUT);
  digitalWrite(messPort,HIGH);

  do
  {
     c_time++;
  }while(!digitalRead(messPort));

  capacitance=(c_time*0.042)*10.0;
```

```
    if(capacitance<999)
    {
        Serial.print(capacitance);
        Serial.println("nF");
    }
    else
    {
        capacitance=capacitance/1000;
        Serial.print(capacitance);
        Serial.println("uF");
    }

    delay(1000);
}
```

10.16 | Reading potentiometers and trimmers the professional way

You already know how to read a potentiometer with the analogRead() command. However, this method is not ideally suited for some tasks, because the last digit of the measurement value jumps up and down. This is caused not only by the given error of the ADC but also by the fact that a potentiometer is relatively imprecise and subject to the so called drift. With the sensitive analog input, we measure this drift as well.

Avoiding inaccuracies in potentiometers

To avoid this, we can again employ a hysteresis function. The value is only updated when the last digit exceeds a given value or drops below it, respectively. Furthermore, the example code outputs the captured value only to Terminal when it is different to the previous one. When you turn the potentiometer or trimmer significantly, a measurement value is displayed in Terminal.

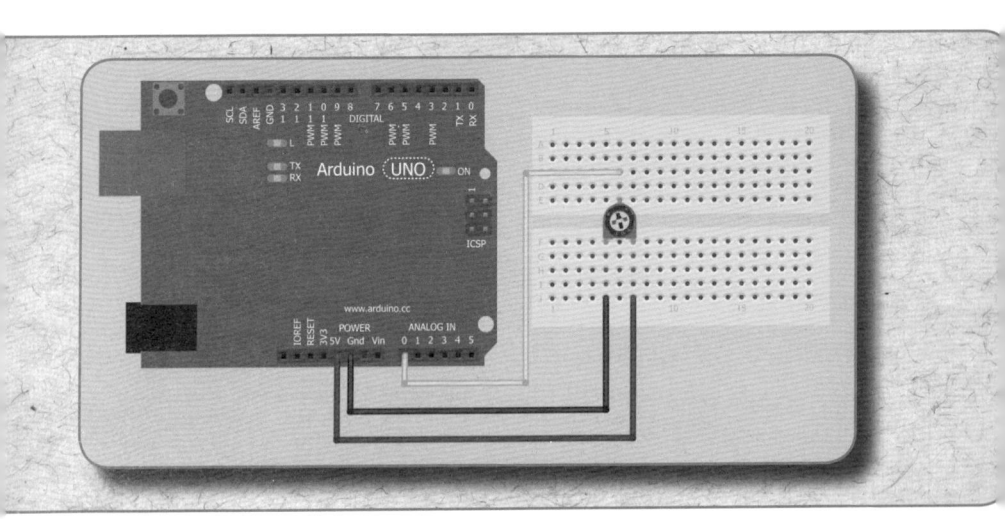

Figure 10.20: Set-up for the experiment; the trimmer is again used in a voltage divider circuit

Required parts for the experiment
- 1 x microcontroller board Arduino Uno
- 1 x breadboard
- 1 x 10 kΩ trimmer
- 3 x jump wire, ca. 10 cm

10 More Experiments with the Arduino 241

Example: *Reading a potentiometer*

Time needed: 10 min
Difficulty: 2

```
// Franzis Arduino
// Reading a potentiometer

int Poti=0;
int raw,raw_last,raw_min,raw_max=0;
int hysterese=10;

void setup()
{
    Serial.begin(9600);
    Serial.println("Reading a potentiometer
                   the professional way");
    Serial.println();
}

void loop()
{
    raw=analogRead(Poti);
    raw_min=raw_last-hysterese;
    raw_max=raw_last+hysterese;
    if((raw!=raw_last))
    {
        if((raw>raw_max)||(raw<raw_min))
        {
            Serial.println(raw);
            raw_last=raw;
        }
    }
}
```

10.17 | State Machines

State machines and their practical usage

A state machine is a software concept modeled after an abstract machine with an internal state. The machine works by going over to another state and executing some actions along the way. The new state develops from the previous one.

The machine is operated by a clock, hence it cannot respond to events in arbitrarily small time periods. In every clock cycle, the current state and the state of the input channels are examined to decide on the next state and the actions to be carried out.

A beverage vending machine as an example for a state machine

Take for example a beverage vending machine. It waits until a customer wants to fetch a beverage. This waiting time we can call state 1. Here, the coin slot is monitored for inserted money. The amount may also be displayed. When a customer inserts money, the machine checks if the inserted money is sufficient to buy a beverage und displays the accrued sum. This is the second state. When enough money is inserted, the machine goes over to state 3.

Now the customer can choose a beverage via the push-buttons. The machine dispenses the product, and the procedure starts again.

Of course, the state machine of a real vending machine is more complex, but this example is sufficient to explain the concept.

10 More Experiments with the Arduino

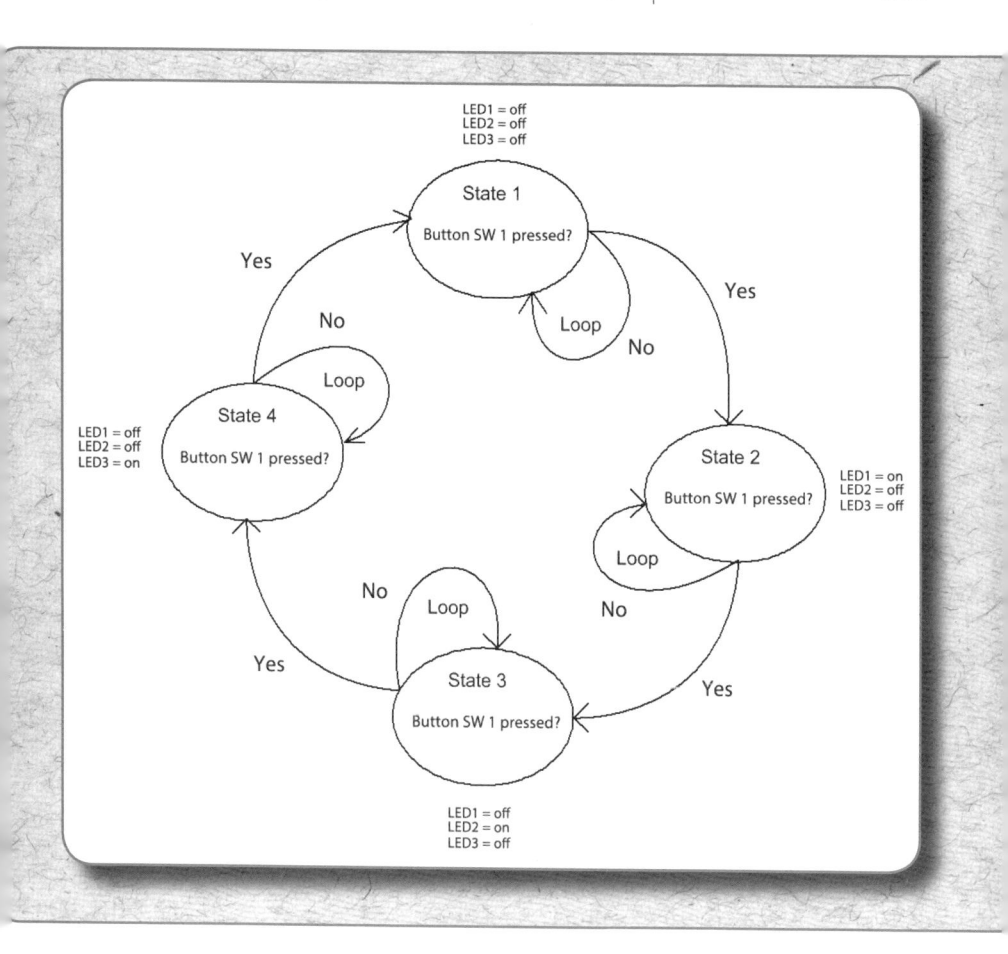

Figure 10.21: A simple state machine that reads push-button SW1 and goes over to the next state depending on the result. In the process, the LEDs 1, 2, and 3 are cycled.

Let us program the state machine shown above that reads the push-button SW1 and jumps forward to the next state whenever the button is pressed.

Programming a simple state machine

In this circuit, you can use any arrangement of differently coloured LEDs you like.

A state machine changing colour

When you press button SW1, the state machine is called and the LEDs switch to the next state. The `Statemachine` function always increments its state by 1 in the `cnt` variable. When the last state is reached, the counter is reset and the whole thing begins again.

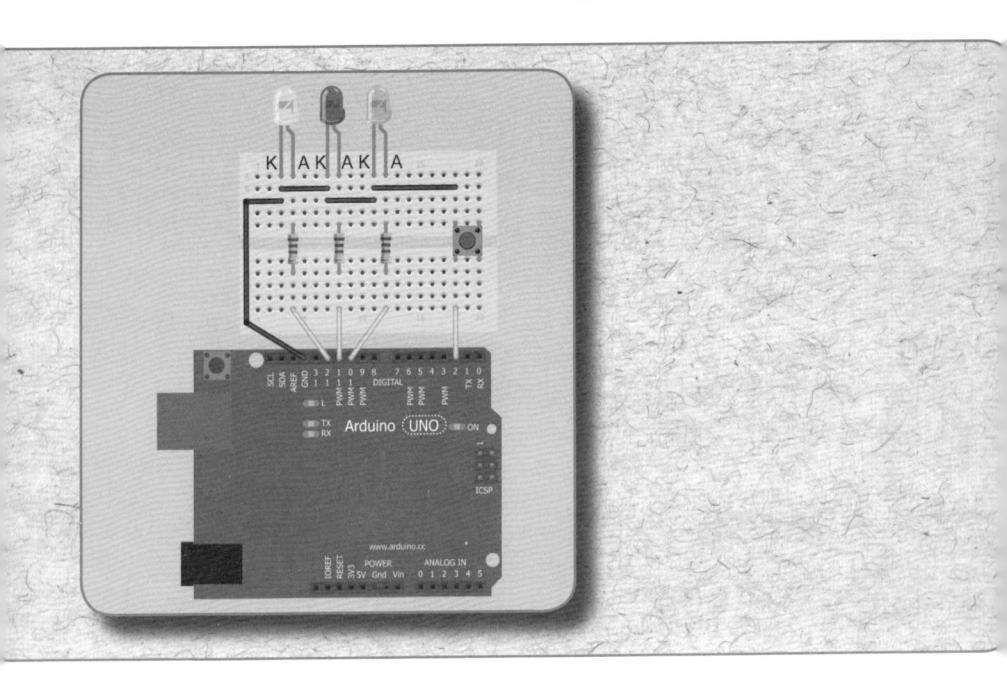

Figure 10.22: Set-up of the state machine

Required parts for the experiment
- 1 x microcontroller board Arduino Uno
- 1 x breadboard
- 1 x push-button
- 1 x red LED
- 1 x yellow LED
- 1 x green LED
- 3 x 1.5 kΩ resistor
- 7 x jump wire, ca. 5 cm
- 1 x jump wire, ca. 10 cm

Example: State machine

Time needed: 10 min
Difficulty: 4

```
// Franzis Arduino
// State machine

int LED1=12;
int LED2=11;
int LED3=10;
int SW1=2;
int cnt=0;
int state=0;

void setup()
{
   pinMode(LED1,OUTPUT);
   pinMode(LED2,OUTPUT);
   pinMode(LED3,OUTPUT);
   pinMode(SW1,INPUT);
   digitalWrite(SW1,HIGH);
}

void loop()
{
   if(!digitalRead(SW1))
   {
      delay(50);
      if(!digitalRead(SW1))
      {
         Statemaschine();
         while(!digitalRead(SW1));
      }
   }
   delay(10);
}
```

```
void Statemaschine(void)
{
   switch(state)
   {
      case 0:
      digitalWrite(LED1,LOW);
      digitalWrite(LED2,LOW);
      digitalWrite(LED3,LOW);
      state++;
      break;

      case 1:
      digitalWrite(LED1,HIGH);
      digitalWrite(LED2,LOW);
      digitalWrite(LED3,LOW);
      state++;
      break;

      case 2:
      digitalWrite(LED1,LOW);
      digitalWrite(LED2,HIGH);
      digitalWrite(LED3,LOW);
      state++;
      break;

      case 3:
      digitalWrite(LED1,LOW);
      digitalWrite(LED2,LOW);
      digitalWrite(LED3,HIGH);
      state=0;
      break;
   }
}
```

10.18 | 6-channel voltmeter

When you want to visualize the measured date on your PC, you can do this with the example shown here. Those who want to edit or extend the PC source code need a copy of Visual Basic Express. You can download it for free at the Microsoft website on:

Simple data transfer from the Arduino

http://www.microsoft.com/visualstudio/eng/downloads

The Arduino program reads the analog inputs 0 to 5 and sends the data to the PC via the UART serial interface. The small PC program receives the data, evaluates the captured ADC values, converts them into Volts, and displays them. This programs demonstrates how to design a simple data transfer solution.

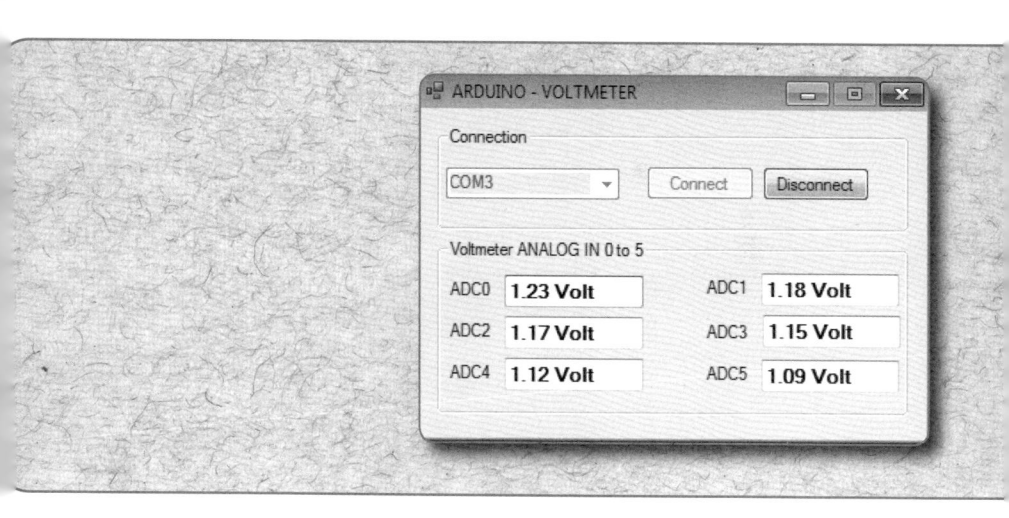

Figure 10.23: 6-channel voltmeter program written in VB.Net

First, the program reads all the ADC channels of th microcontroller and stores the captured data as intege variables in the `Adc_raw()` array. In the next step, th two-byte variables in the array are decomposed into *high* and a *low* byte. Then, the high byte is sent to the PC followed by the low byte. To check the data packet for th constant is XORed with the high and the low byte to yiel a checksum. This checksum is also sent to the PC. The pro cess is repeated until the readings of all six channels ar transferred to the PC.

Building blocks of the program

The program on the PC reads the bytes and store them in a data array. Now the high and the low bytes ar merged again to recreate the original ADC value. The P program then calculates the checksum in the same wa as the Arduino program did, but this time with the dat received on the PC. If the received checksum matche the calculated one, the measured value is displayed i the text boxes (`TextBoxEx`). Otherwise, the value is no updated and the program waits for the next data transfe

Example: Voltmeter

Time needed: 10 min
Difficulty: 3

```
// Franzis Arduino
// 6-channel voltmeter

int LED=13;
char startbyte=0;
int highbyte=0;
int lowbyte=0;
int adc_raw[6];
int adc_cnt=0;
int cnt=0;
int crc=0;
```

10 More Experiments with the Arduino

```
void setup()
{
  Serial.begin(9600);
  pinMode(LED,OUTPUT);
}

void loop()
{
  startbyte=Serial.read();
  if(startbyte==42)
  {
    digitalWrite(LED,HIGH);
    delay(50);
    digitalWrite(LED,LOW);
    delay(50);

    Serial.flush();

    for(cnt=0;cnt<6;cnt++)
    {
      adc_raw[cnt]=analogRead(adc_cnt);
      adc_cnt++;
    }
    adc_cnt=0;

    for(cnt=0;cnt<6;cnt++)
    {
      highbyte=adc_raw[cnt]/256;
      lowbyte=adc_raw[cnt]%256;
      Serial.write(highbyte);
      Serial.write(lowbyte);
    }

    crc=170^highbyte^lowbyte;
    Serial.write(crc);
  }
}
```

10.19 | Programming Your Own Voltage Plotter

The analog input as analog data recorder

In this example, we will use the analog input as an analog data recorder. The data transfer is implemented in the same way as in the previous program (the 6-channel voltmeter). The captured ADC value is stored in an integer variable and again decomposed into a high and a low byte. These bytes are send to the PC via the UART interface.

The values of the high and the low byte are calculated as follows:

```
High byte = raw value / 256
Low byte = raw value % 256
```

To calculate the high byte, we divide the captured ADC value by 256 (which corresponds to one byte). The result indicates how often the raw value fits into 256. To get the remainder, we carry out the modulo operation (%). This yields the lower eight bits (the least significant byte).

Look at the following example:

The value 766 has to be decomposed into a high and a low byte. According to the formulas given above, we get the following result:

```
High byte = 2
Low byte = 254
```

To recreate the number 766 from these bytes, we have to multiply the high byte by 256 and add the low byte 254.

10 More Experiments with the Arduino 251

The voltage plotter program records the measured values in intervals of 100 ms, 500 ms, and 1000 ms. In the desired intervals, the program on the PC sends a start byte (55) to the microcontroller. When the Arduino receives this byte, it commences measuring. Then the captured value and the checksum are sent back to the PC program that recalculates the original value from the two bytes and displays it.

The voltage plotter program

This form of data recording can be used for many different applications. Some good examples include:

Possible usages for the voltage plotter

- Temperature recording
- Voltage gradient when charging or discharging
- Voltage monitoring

The source code of the program can be found on the accompanying CD.

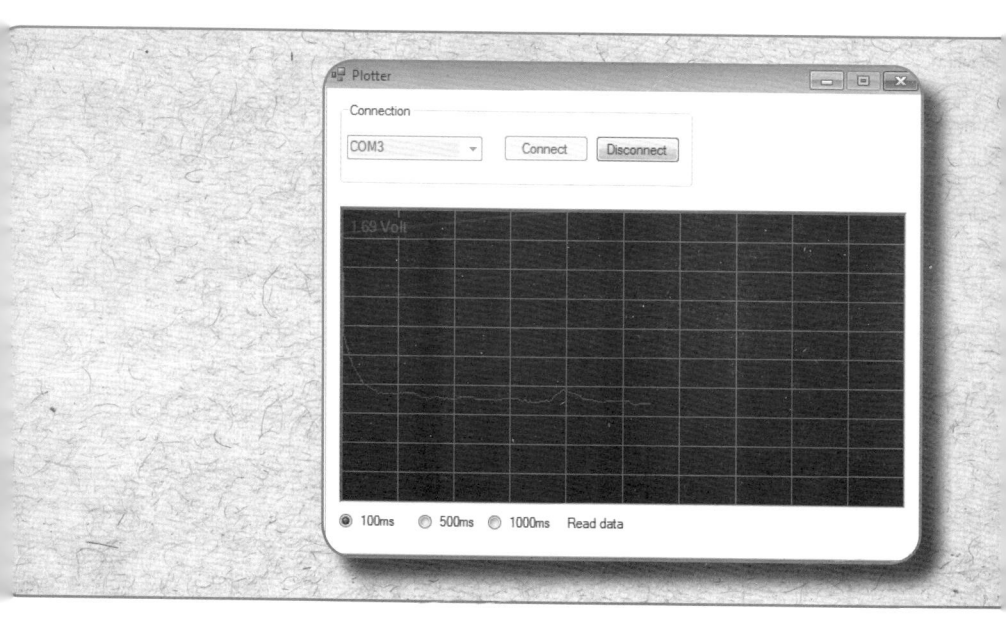

Figure 10.24: *The voltage plotter program on the PC*

A lightness plotter for your home

Tip You can build a little lightness plotter using an LDR. Attach the anode to analog input 0 on the pin header and the cathode to ground. When you now cast light on the LDR or shade it, the voltage changes. You can then try to scale the amplitude in the plotter (the »height« of the voltage graph) according to the actual LDR.

Example: *Voltage plotter*

Time needed: 10 min
Difficulty: 3

■ ■ ■ ☐ ☐

```
// Franzis Arduino
// Voltage plotter

char startbyte=0;
int highbyte=0;
int lowbyte=0;
int adc=0;
int crc=0;

void setup()
{
   Serial.begin(9600);
}

void loop()
{
   startbyte=Serial.read();
   if(startbyte==55)
   {
      Serial.flush();
      adc=analogRead(0);
      highbyte=adc/256;
      lowbyte=adc%256;
      Serial.write(highbyte);
      Serial.write(lowbyte);
      crc=170^highbyte^lowbyte;
      Serial.write(crc);
   }
}
```

10.20 | Arduino Storage Oscilloscope

If you want to measure fast signals up to ca. 5 kHz you will have to record them in another way than you did in the plotter program. First, we create an array that can hold 256 measuring values. Then, we have a loop that counts from 0 to 255. In every pass, it measures the analog input and writes the value into the array. Measurement always begins when the start byte (55) is sent from the PC. After completing the measurement, the array is transferred to the VB.Net program on the PC. The decomposition into the high and the low byte takes place as before.

The Arduino storage oscilloscope can capture fast signals up to 5 kHz

The PC program evaluates the data and draws the voltage development into the graphics section. This way you get a small storage oscilloscope that allows you to analyze low-frequency signals. If you want to measure voltages of more than 5 V DC you will have to add a voltage divider to the analog input.

When you have no appropriate signal source (waveform generator) you can visualize the 50 Hz hum. Attach a piece of wire to the analog input 0 and hold it in the proximity of electronic devices, e.g. a PC monitor. You can also try to approach the wire with your hands. Look how the oscillogram changes!

The analog input has such a high resistance that even smallest changes in voltage become visible. Do you happen to own an old decommissioned dynamic microphone lounging around in your workshop? When you attach it to the oscilloscope it will clearly display the oscillations of the microphone coil. This also works well with a sound converter and can uncover even the smallest vibrations.

Attaching a dynamic microphone will display the oscillations of the microphone coil

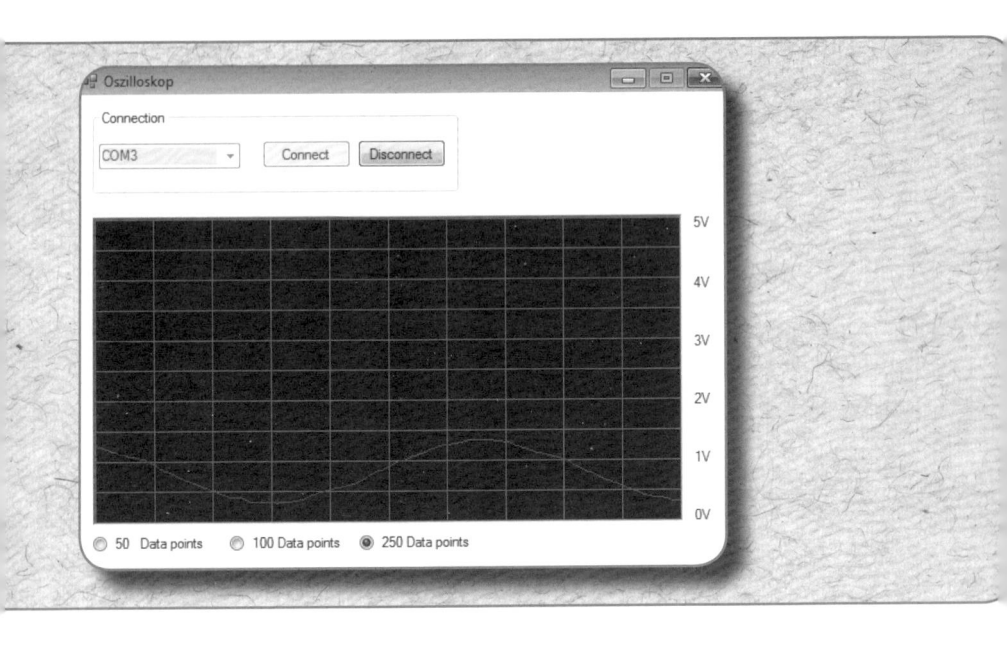

Figure 10.25: The storage oscilloscope program on the PC

Example: Oscilloscope

Time needed: 10 min
Difficulty: 3

```
// Franzis Arduino
// Oscilloscope

char startbyte=0;
int highbyte=0;
int lowbyte=0;
int adc[256];
int cnt=0;
int crc=0;

void setup()
{
    Serial.begin(115200);
}
```

```
void loop()
{
    startbyte=Serial.read();

    if(startbyte==55)
    {
        Serial.flush();

        for(cnt=0;cnt<256;cnt++)
        {
            adc[cnt]=analogRead(0);
        }

        for(cnt=0;cnt<256;cnt++)
        {
            highbyte=adc[cnt]/256;
            lowbyte=adc[cnt]%256;
            Serial.write(highbyte);
            Serial.write(lowbyte);
        }
        crc=170^highbyte^lowbyte;
        Serial.write(crc);
    }
}
```

10.21 | StampPlot: a professional data logger – free of charge!

StampPlot is a program to plot, display, log, and monitor serial data originating from a microcontroller. In the previous examples, you have seen how to implement data recording applications (oscilloscope, voltage plotter, etc.)

StampPlot has more to offer!

by means of VB.Net programs. StampPlot takes this to much higher level. It offers countless functions for recording data and interpreting measured values.

In our example, you capture the voltage development at analog input 0, where some device is attached, say potentiometer. Before you can do that, you first have to install the StampPlot program.

You can download it at:

http://www.selmaware.com/

After installation, start the program and click one of the preset display options.

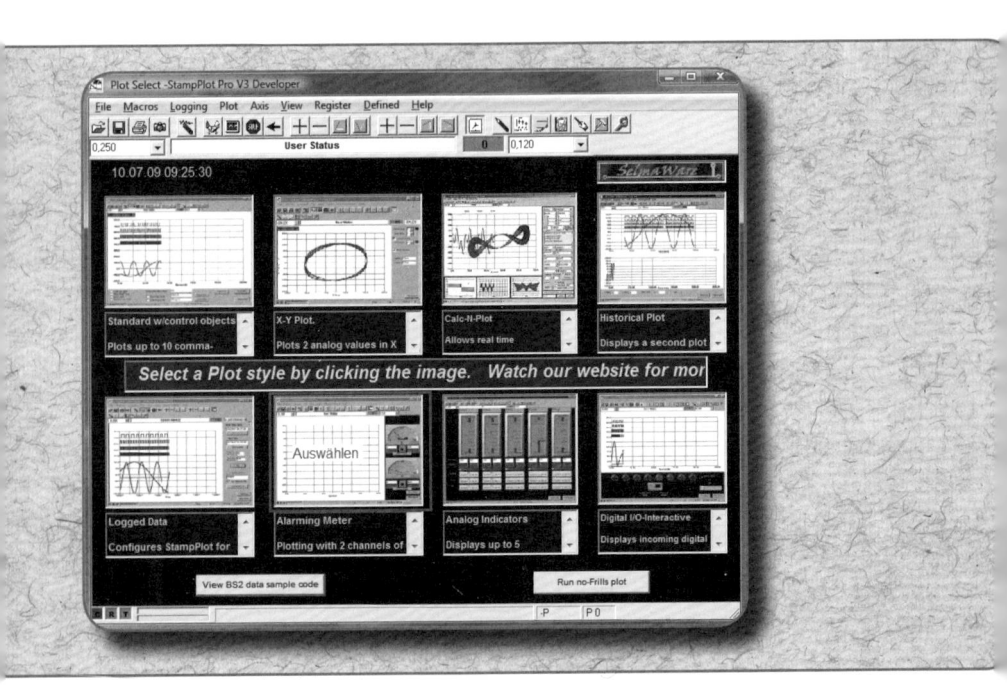

Figure 10.26: Selecting the plot style in StampPlot

10   More Experiments with the Arduino   257

After selecting the graph type, you have to configure the serial interface to which the Arduino board is attached. The board contains a USB-to-UART bridging chip that provides a virtual COM port. Hence, you just have to select the same COM port that you defined in the Arduino IDE for programming and for the Terminal output.

Configuring the serial interface

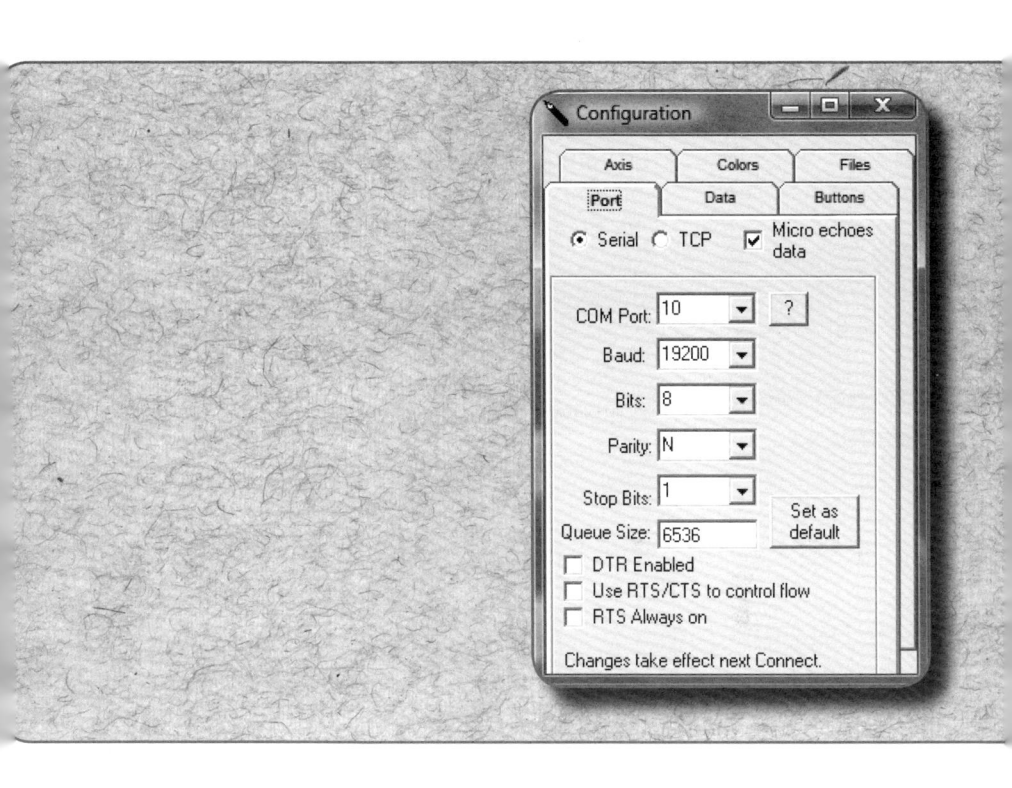

Figure 10.27: Configuring StampPlot

After uploading the Arduino program to the microcontroller, you have ten seconds to connect StampPlot with the serial interface. If that time span is not sufficient, you can press reset.

Connecting StampPlot with the serial interface

When StampPlot is connected to the microcontroller, the recording begins ten seconds after starting the program on the Arduino. The first information sent by the microcontroller tells StampPlot how to display the recording. This includes settings like graph reset, data points, resolution, etc.

Measuring data and transferring them to StampPlot

After that, the data are measured and transferred to StampPlot. If you want to know more about StampPlot, you can have a look at the Selmaware website and study the manual. StampPlot is a very comprehensive program and you could write a tome of its own about it. This little introduction is just intended to facilitate your first steps.

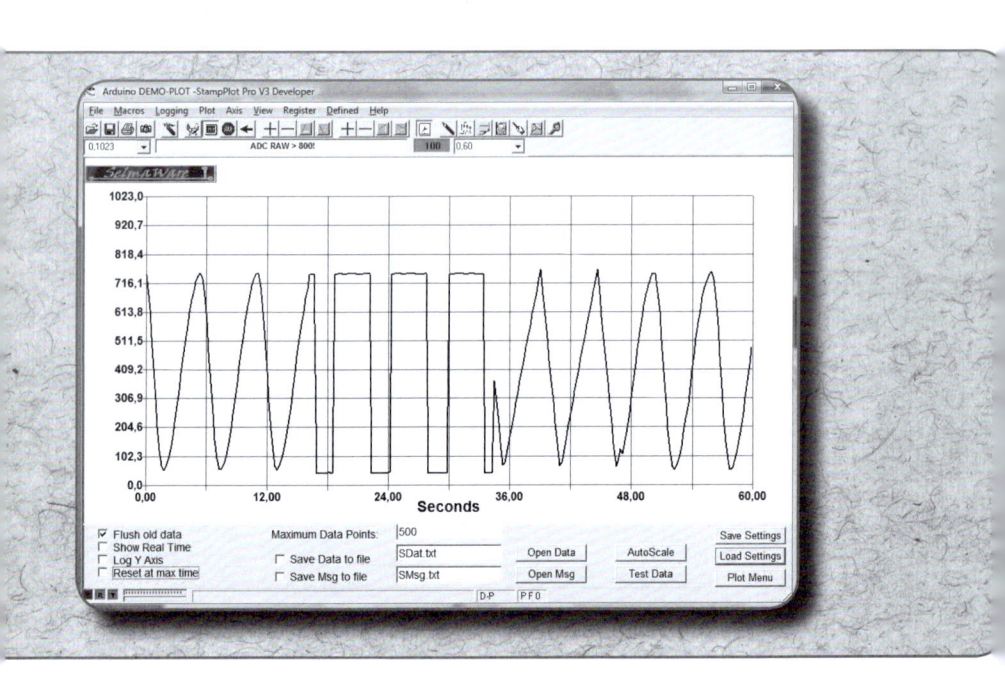

Figure 10.28: Capturing values in StampPlot

Example: StampPlot

Time needed: 10 min
Difficulty: 4

■ ■ ■ ■ □

```
// Franzis Arduino
// StampPlot data logger

int LED=13;
int adc0=0;

void setup()
{
   Serial.begin(19200);
   pinMode(LED,OUTPUT);

   delay(10000);

   // Send StampPlot settings
   Serial.println("!RSET");
   // Reset plot to clear data
   Serial.println("!TITL Arduino DEMO-PLOT");
   // Caption form
   Serial.println("!PNTS 300");
   // 1000 sample data points
   Serial.println("!TMAX 60");
   // Max. 60 seconds
   Serial.println("!SPAN 0,1023");
   // 0-1023 span
   Serial.println("!AMUL 1");
   // Multiply data by 1
   Serial.println("!DELD");
   // Delete data file
   Serial.println("!SAVD ON");
   // Save data
   Serial.println("!TSMP ON");
   // Time stamp on
   Serial.println("!CLMM");
```

```
    // Clear min/max
    Serial.println("!CLRM");
    // Clear messages
    Serial.println("!PLOT ON");
    // Start plotting
    Serial.println("!RSET");
    // Reset plot to time 0
}

void loop()
{
    adc0=analogRead(0);
    Serial.print(adc0);
    Serial.write(13);

    if(adc0>700)
    {
        Serial.println("!USRS ADC RAW > 800!");
        Serial.println("ADC RAW is greater than 800");
        digitalWrite(LED,HIGH);
    }

    if(adc0<150)
    {
        Serial.println("!USRS ADC RAW < 250!");
        Serial.println("ADC RAW is smaller than 150");
        digitalWrite(LED,LOW);
    }

    delay(200);
}
```

0.22 | Controlling the Arduino Pins via the Arduino Ports Program

This section explains how to send data from the PC to the microcontroller to evaluate them there. The Arduino Ports program sends a single byte to the controller. With one byte, you could control 255 digital outputs or change an analog output.

Sending data from the PC to the microcontroller and evaluating them

The example is consciously kept simple, and there are no checksums involved in the data transfer. This is again a good exercise for you! Try to modify the program so that you can change the brightness of an LED via the analog output and control the remaining I/O pins. Of course you should also provide a checksum test (CRC) as we have done it in the data plotter and the oscilloscope.

In the following experiment, you turn on and off LEDs via the Arduino Ports program. Use the *LED* buttons for the LED L on the Arduino Uno and the *DIGITAL 5* buttons for the external LED on the breadboard. (To turn them on, click the *ON* button, and to turn them off, click *OFF*.) The program and its source code can be found on the accompanying CD.

Figure 10.29: The program to control the digital Arduino pins

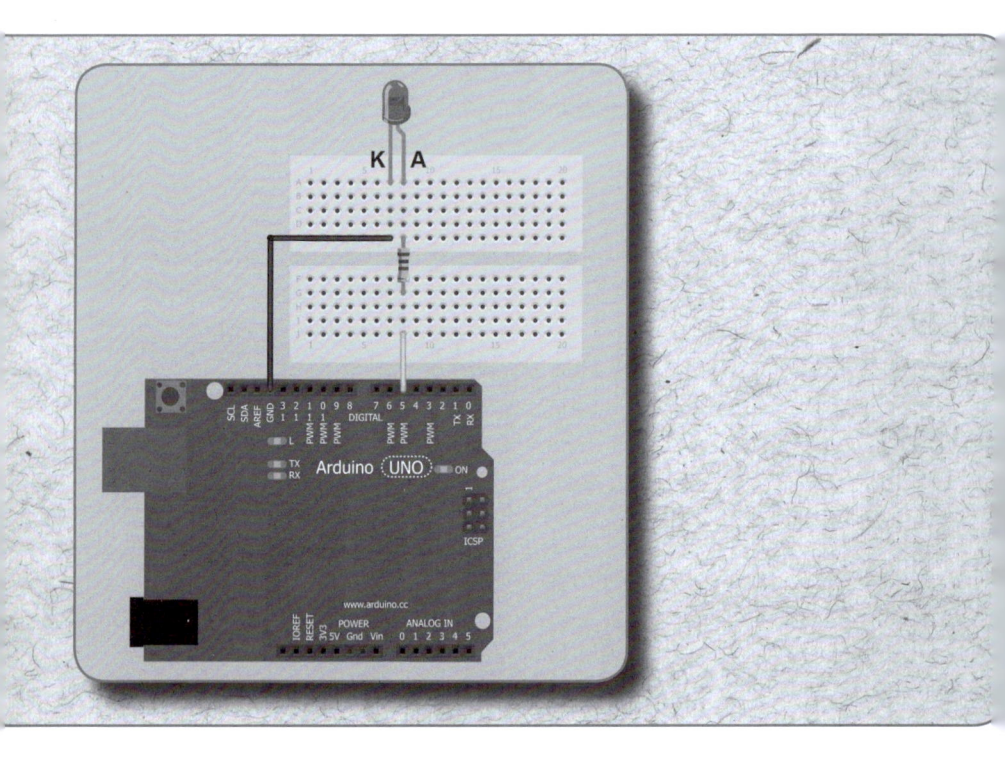

Figure 10.30: Set-up for the experiment

10 More Experiments with the Arduino

- 1 x microcontroller board Arduino Uno
- 1 x breadboard
- 1 x red LED
- 1 x 1.5 kΩ resistor
- 1 x jump wire, ca. 5 cm
- 1 x jump wire, ca. 10 cm

Required parts for the experiment

xample: Controlling Arduino pins

```
// Franzis Arduino
// Controlling pins via the Arduino Ports program

int LED=13;
int IO_5=5;
int input=0;

void setup()
{
    Serial.begin(9600);
    pinMode(LED,OUTPUT);
    pinMode(IO_5,OUTPUT);
}

void loop()
{
    input=Serial.read();
    switch(input)
    {
        case 10:
        digitalWrite(LED,HIGH);
        break;
```

Time needed: 10 min
Difficulty: 3

```
        case 20:
        digitalWrite(LED,LOW);
        break;

        case 30:
        digitalWrite(IO_5,HIGH);
        break;

        case 40:
        digitalWrite(IO_5,LOW);
        break;

        case 100:
        digitalWrite(LED,LOW);
        digitalWrite(IO_5,LOW);
        break;
    }
}
```

10.23 | Temperature Switch

How to build a temperature switch

This experiment demonstrates how to build a temperature switch using a small silicon diode. The forward voltage V_F of a diode depends on the ambient temperature. The higher the temperature, the lower V_F and vice versa. This voltage can be measured at an ADC input. This way you can implement a temperature switch.

However, the variation is quite small. The 10-bit ADC of the Arduino microcontroller is only suited for a temperature switch. If you use a measuring amplifier (operational amplifier), you can spread the range and build a thermometer.

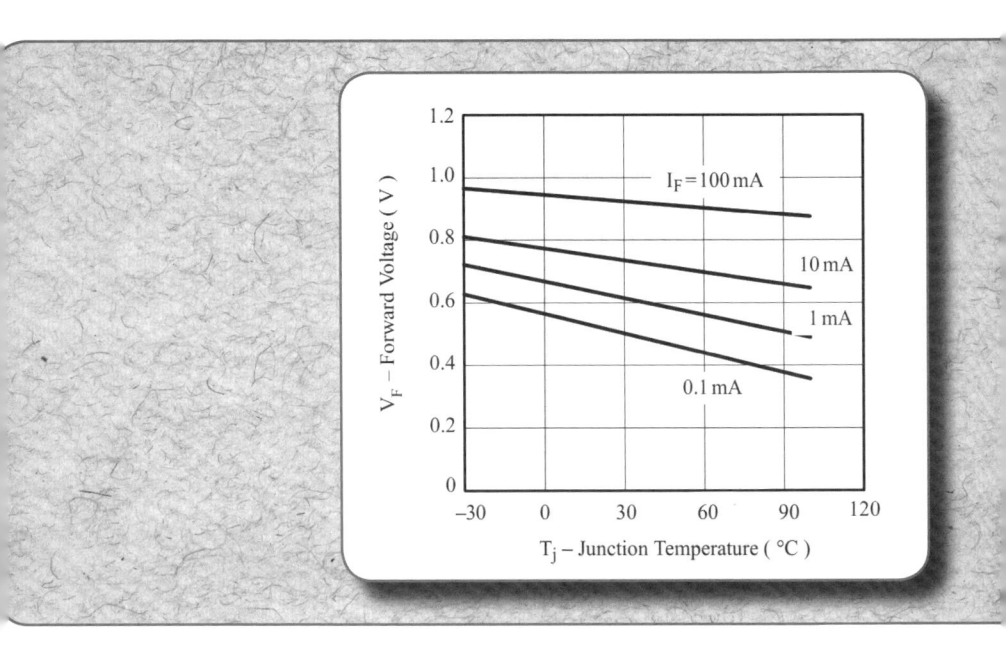

Figure 10.31: The relationship between ambient temperature and forward voltage V_F at given currents (Source: Visay data sheet)

The example program turns on LED L once the temperature drops below a preset threshold. After that, you can carefully warm up the diode with a lighter. Eventually, the LED will turn off again.

Temperature thresholds

- 1 x microcontroller board Arduino Uno
- 1 x breadboard
- 1 x diode 1N4148
- 1 x 47 kΩ resistor
- 1 x jump wire, ca. 5 cm
- 2 x jump wire, ca. 10 cm

Required parts for the experiment

Figure 10.32: Set-up of the temperature switch with the 1N4148 diode

Example: *Temperature switch*

```
// Franzis Arduino
// Temperature switch

int LED=13;
int Uf=0;

void setup()
{
    Serial.begin(9600);
    pinMode(LED,OUTPUT);
}

void loop()
{
    Uf=analogRead(0);
    Serial.print("Uf = ");
    Serial.println(Uf);

    if(Uf>40)digitalWrite(LED,HIGH);
    if(Uf<20)digitalWrite(LED,LOW);

    delay(250);
}
```

Time needed: 10 min
Difficulty: 3
■ ■ ■ ☐ ☐

The forward voltage V_F is output to Terminal. Now you can read the voltage at room temperature. You can adjust and even reverse the thresholds as you like it. This way, you can construct a temperature-controlled fan control.

TURN ON YOUR CREATIVITY

FRANZIS
ARDUINO

THE
FRITZING
PROGRAM

**...FURTHER
TOOLS
FOR YOUR
EVERYDAY
PROGRAMMING**

11

Creating circuit diagrams with Fritzing

You may have asked yourself how the illustrations in this book were made. The program used for the drawings is called *Fritzing* and can be downloaded free of charge at:

http://fritzing.org

With this program you can easily design circuits as you have seen in the examples in this book. You can also draw circuit diagrams and layouts. Visiting the website is worthwhile, as it contains useful information on the Arduino.

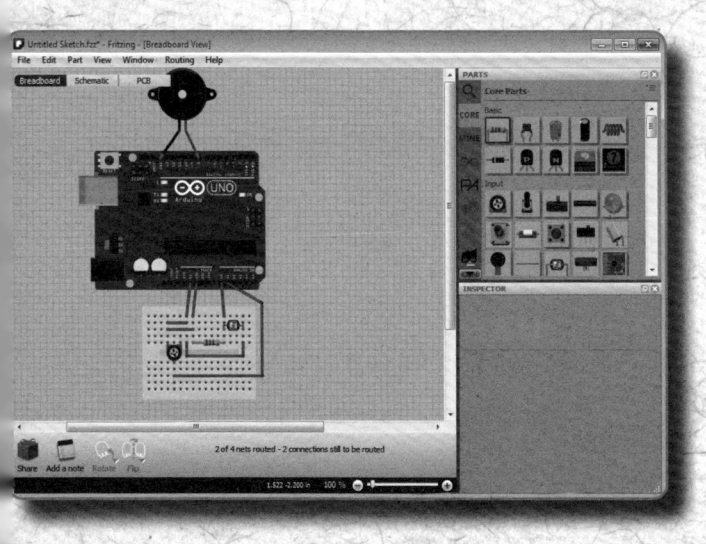

Figure 11.1: The Fritzing program in action

TURN ON YOUR CREATIVITY

FRANZIS
ARDUINO

THE
PROCESSING
PROGRAM

... PROCESSING –
THE ICING ON THE
PROGRAMMING CAKE

12

Processing: the powerful development program for the PC

To complete our discussion, we will introduce another interesting PC program called Processing. It was mentioned briefly in the previous chapters and may be regarded as an ancestor of the Arduino IDE. Processing is extremely powerful and nearly as simple to use as the Arduino IDE to program the Arduino board.

Many examples are provided to facilitate the first steps with Processing. It is possible not only to implement simple programs that communicate with the Arduino via the serial interface but to calculate complex 3D graphics as well. The program offers many possibilities and is platform independent.

Figure 12.1: Splash screen of Processing

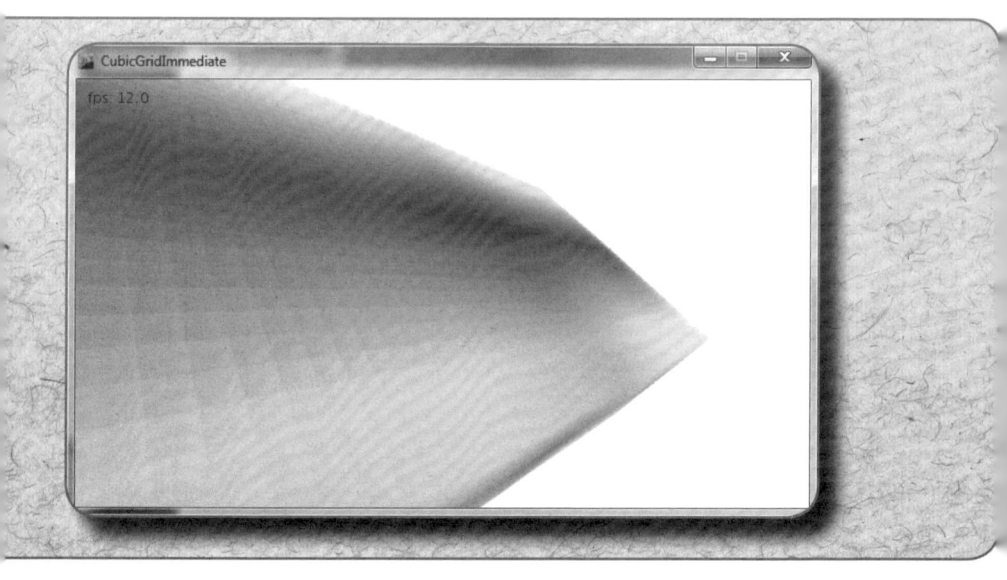

Figure 12.2: A cubic 3D animation in Processing

12 The Processing Program | 273

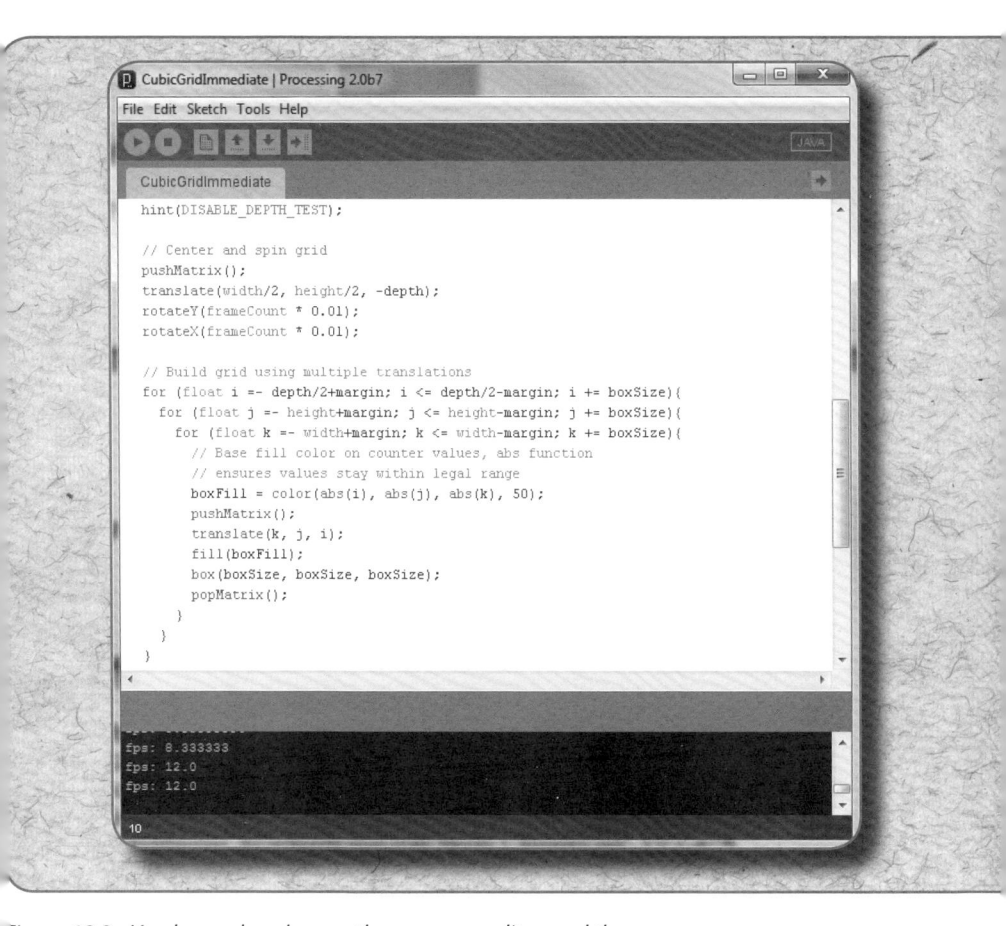

Figure 12.3: You have already seen the program editor and the basic commands in the Arduino IDE. Therefore, it should be no problem to become familiar with Processing.

Have fun and success with the Arduino!

TURN ON YOUR CREATIVITY

FRANZIS
ARDUINO

APPENDIX

...ABBREVIATIONS, QUANTITIES, AND UNITS

13

On the following pages, you will find some useful tables for abbreviations, electrical quantities, units of measurements and symbols.

13.1 | Electrical quantities

You have to differentiate between quantities like voltage, current, and resistance and the units of measurement for these quantities (volt, ampere, and ohm). Every quantity and every unit of measurement has its own abbreviation that is used in formulas. This provides for a neat and clear notation. For instance, you simply write »I = 1 A« instead of »current = 1 ampere«.

In this book, the following abbreviations are used:

Quantity	Abbreviation	Unit	Abbreviation
Voltage	V or U	Volt	V
Current	I	Ampere	A
Resistance	R	Ohm	Ω
Power	P	Watt	W
Frequency	f	Hertz	Hz
Time	t	Second	s
Wave length	λ (lambda)	Meter	m
Inductance	L	Henry	H
Capacitance	C	Farad	F
Area	A	Square meter	m^2

13.2 | ASCII Table

Symbol	Decimal	Hexa-decimal	Binary	Description
NUL	000	000	00000000	Null character
SOH	001	001	00000001	Start of header
STX	002	002	00000010	Start of text
ETX	003	003	00000011	End of text
EOT	004	004	00000100	End of transmission
ENQ	005	005	00000101	Enquiry
ACK	006	006	00000110	Acknowledgment
BEL	007	007	00000111	Bell
BS	008	008	00001000	Backspace
HAT	009	009	00001001	Horizontal tab
LF	010	00A	00001010	Line feed
VT	011	00B	00001011	Vertical tab
FF	012	00C	00001100	Form feed
CR	013	00D	00001101	Carriage return
SO	014	00E	00001110	Shift out
SI	015	00F	00001111	Shift in
DLE	016	010	00010000	Data link escape
DC1	017	011	00010001	Device control 1
DC2	018	012	00010010	Device control 2
DC3	019	013	00010011	Device control 3
DC4	020	014	00010100	Device control 4
NAK	021	015	00010101	Negative acknowledgment
SYN	022	016	00010110	Synchronous idle
ETB	023	017	00010111	End of transmission block
CAN	024	018	00011000	Cancel

Symbol	Decimal	Hexa-decimal	Binary	Description
EM	025	019	00011001	End of medium
SUB	026	01A	00011010	Substitute
ESC	027	01B	00011011	Escape
FS	028	01C	00011100	File separator
GS	029	01D	00011101	Group separator
RS	030	01E	00011110	Request to send, record separator
US	031	01F	00011111	Unit separator
SP	032	020	00100000	Space
!	033	021	00100001	Exclamation mark
"	034	022	00100010	Double quote
#	035	023	00100011	Number sign
$	036	024	00100100	Dollar sign
%	037	025	00100101	Percent
&	038	026	00100110	Ampersand
'	039	027	00100111	Single quote
(	040	028	00101000	Left opening parenthesis
)	041	029	00101001	Right closing parenthesis
*	042	02A	00101010	Asterisk
+	043	02B	00101011	Plus
,	044	02C	00101100	Comma
-	045	02D	00101101	Minus or dash
.	046	02E	00101110	Dot
CHAR	DEC	HEX	BIN	Description
/	047	02F	00101111	Forward slash
0	048	030	00110000	
1	049	031	00110001	
2	050	032	00110010	

Symbol	Decimal	Hexa-decimal	Binary	Description
3	051	033	00110011	
4	052	034	00110100	
5	053	035	00110101	
6	054	036	00110110	
7	055	037	00110111	
8	056	038	00111000	
9	057	039	00111001	
	058	03A	00111010	Colon
	059	03B	00111011	Semicolon
<	060	03C	00111100	Less than
=	061	03D	00111101	Equal
>	062	03E	00111110	Greater than
?	063	03F	00111111	Question mark
@	064	040	01000000	At symbol
A	065	041	01000001	
B	066	042	01000010	
C	067	043	01000011	
D	068	044	01000100	
E	069	045	01000101	
F	070	046	01000110	
G	071	047	01000111	
H	072	048	01001000	
	073	049	01001001	
	074	04A	01001010	
	075	04B	01001011	
	076	04C	01001100	
M	077	04D	01001101	
N	078	04E	01001110	

Symbol	Decimal	Hexa-decimal	Binary	Description
O	079	04F	01001111	
P	080	050	01010000	
Q	081	051	01010001	
R	082	052	01010010	
S	083	053	01010011	
T	084	054	01010100	
U	085	055	01010101	
V	086	056	01010110	
W	087	057	01010111	
X	088	058	01011000	
Y	089	059	01011001	
Z	090	05A	01011010	
[	091	05B	01011011	Left opening bracket
\	092	05C	01011100	Backslash
]	093	05D	01011101	Right closing bracket
^	094	05E	01011110	Caret
CHAR	DEC	HEX	BIN	Description
_	095	05F	01011111	Underscore
`	096	060	01100000	
a	097	061	01100001	
b	098	062	01100010	
c	099	063	01100011	
d	100	064	01100100	
e	101	065	01100101	
f	102	066	01100110	
g	103	067	01100111	

13 Appendix

Symbol	Decimal	Hexa-decimal	Binary	Description
h	104	068	01101000	
i	105	069	01101001	
j	106	06A	01101010	
k	107	06B	01101011	
l	108	06C	01101100	
m	109	06D	01101101	
n	110	06E	01101110	
o	111	06F	01101111	
p	112	070	01110000	
q	113	071	01110001	
r	114	072	01110010	
s	115	073	01110011	
t	116	074	01110100	
u	117	075	01110101	
v	118	076	01110110	
w	119	077	01110111	
x	120	078	01111000	
y	121	079	01111001	
z	122	07A	01111010	
{	123	07B	01111011	Left opening brace
		124	07C	01111100	Vertical bar
}	125	07D	01111101	Right closing brace
~	126	07E	01111110	Tilde
DEL	127	07F	01111111	Delete